Springer Series in Optical Sciences   Volume 51

Edited by Arthur L. Schawlow

Springer Series in Optical Sciences

Tunable
Solid State Lasers for
Remote Sensing

Proceedings of the NASA Conference
Stanford University, Stanford, USA, October 1–3, 1984

Editors:
R. L. Byer, E. K. Gustafson, and R. Trebino

With 58 Figures

Springer-Verlag Berlin Heidelberg GmbH

Professor Robert L. Byer, Ph. D.
W. W. Hansen Laboratories of Physics, Stanford University, Stanford, CA 94305, USA

Eric K. Gustafson, Ph. D.
Department of Physics, University of California, Berkeley, CA 94720, USA

Rick Trebino, Ph. D.
Department of Applied Physics, Stanford University, Stanford, CA 94305, USA

ISBN 978-3-662-13561-7   ISBN 978-3-540-39765-6 (eBook)
DOI 10.1007/978-3-540-39765-6

Preface

The Workshop on Tunable Solid State Lasers for Remote Sensing was held at Stanford University in October 1984 to assess the state of the art in tunable solid state lasers for remote sensing from satellite platforms. The value of conducting global remote sensing measurements of atmospheric chemistry, climate, and weather in the 1990s is now established. What is not yet defined, however, is the status of the developing tunable laser technology that must meet both the scientific requirements and the space platform constraints. This workshop was convened by the Office of Aeronautics and Space Technology (OAST) of the National Aeronautics and Space Administration (NASA) to assess the status and progress in tunable solid state laser sources for remote sensing.

The workshop was organized to facilitate information exchange across a number of technologies from remote sensing requirements to crystal growth of the materials important for the development of the tunable laser sources. The emphasis was on the recent developments in tunable solid state laser sources necessary to meet the future transmitter requirements for global remote sensing. A goal of the workshop was to form recommendations to NASA on the current and future prospects for solid state laser technology that will allow remote sensing measurements from air, shuttle, and free-flying satellite platforms. The emphasis was on solid state laser sources because they offer the best potential for meeting the demanding requirements of compact size, good efficiency, and long operational lifetimes required for future space station and free-flying platform operation.

The relatively mature Alexandrite tunable laser system was reviewed with emphasis on operational stability, wavelength tunability, and line-width control. The emerging tunable laser system based on the titanium ion doped into sapphire was discussed because of its wide tuning range in the 700–900 nm range which is the wavelength region of interest for humidity and pressure measurements. The review of tunable laser systems included a discussion of the F-center laser systems because of recent progress in the operation of oxide F-center laser sources. The workshop witnessed the announcement of a new laser – the F-center laser in diamond reported by S.C. Rand and L.G. DeShazer.

Progress in lead-salt diode lasers was reviewed by Wayne Lo of the Physics Department of the General Motors Research Laboratories. It is with sadness that we learned of Wayne Lo's death only a few months following the workshop.

The workshop included a panel discussion of the very new and potentially significant technology of diode laser pumped solid state laser sources. This developing technology has the potential to meet the demands for high efficiency and extreme lifetimes that are required for future free-flying satellite applications. The projected time and cost to full development of diode laser arrays for pumping solid state laser sources was debated at length in the panel session.

The measurement of wind requires coherent transmitter-receiver capability. Only recently has the Nd:YAG laser source demonstrated the linewidth required for coherent LIDAR applications. The all-solid-state nature of coherent LIDAR at 300 THz, based on the well developed Nd:YAG technology, is attractive. However, applications may be hampered by the concern for eye-safe operation.

The session on crysal growth emphasized the limitations that material development has placed on the solid state laser source performance and on the efficient nonlinear conversion of existing laser sources. The workshop discussion led to the recognition that materials development is critical to the eventual success of the solid state laser systems in remote sensing applications.

The workshop followed the first International Conference on Tunable Solid State Lasers held in La Jolla, CA in June, 1984. The second Conference on Tunable Solid State Lasers was held in conjunction with the CLEO conference in May 1985. The third conference is planned again to be held just prior to the joint IQEC-CLEO conference in June of 1986. This NASA workshop addressed the topic of tunable solid state lasers with a broader perspective than the above conferences and included discussions of the scaling of solid state lasers to higher peak and average power levels, crystal growth progress, and nonlinear frequency extension techniques. The workshop thus complemented the Tunable Solid State Laser Conferences. The resurgence of interest in solid state laser sources comes nearly a quarter of a century after the demonstration of the first laser by Maiman.

The workshop was organized by Robert L. Byer. The program committee consisted of E.V. Browell, L.G. DeShazer, E.D. Hinkley, E.T. Menzies, and P.F. Moulton. The contributions of the program committee and R. Curran and F. Allario of NASA led to an outstanding program. The success of the workshop was due to the administrative and support efforts of the workshop treasurer, R.C. Eckhardt, and the workshop secretary, Mary Farley. The list of attendees was typed by Douglas R. Byer.

All sessions of the workshop were video taped. The video tapes have been edited and are available through the Stanford Laser Consulting Group, 835 Webster St., Suite E, Palo Alto, CA 94301. The five presentations listed below were given at the workshop but are not included as written manuscripts in this book. However, transcripts with figures are available for the presentations by John Emmett and Peter Moulton.

Solid State Laser Performance – An Overview	By John Emmett
Review of Tunable Solid State Laser Systems	By Peter Moulton
Progress in Titanium Sapphire	By Peter Moulton
High Average Power Nd:GSGG Studies	By Peter Moulton
Progress in Diode Array Pumped Nd:YAG	By R. Rice

Stanford, USA   *R. L. Byer*
October 1, 1985   *E. K. Gustafson · R. Trebino*

Contents

Introduction

R.L. Byer

Department of Applied Physics, Stanford University,
Stanford, CA 94305, USA

The goal of the NASA Workshop on Tunable Solid State Lasers for Remote Sensing is to evaluate solid state laser technology for future remote sensing applications from aircraft, from space shuttle, and from free-flying space platforms. The solid state laser technology is of interest because of its potential for small volume, lightweight, and for efficient, long lifetime operation. One goal of the workshop is to determine just how close we are to realizing that potential and to determine what needs to be done to achieve it.

The workshop was organized to bring together representatives from NASA, the sponsoring agency, from the remote sensing community, from the laser engineering community, from the laser materials community and from the nonlinear optics community. The goal is to promote discussions within and among these groups to determine answers to the questions of tunable solid state laser development in way that can be cast into recommendations to NASA. The recommendations need to be made in terms of the time and resources required, and the associated risks, for the development of solid state laser technology for the remote sensing applications of interest to NASA.

The guest speaker for the working dinner--a NASA euphemism for a banquet-- is Professor Peter Banks of the Electrical Engineering Department of Stanford Univesity. Professor Banks will talk on "Space Science in the Space Station Era". His talk will review discoveries of the NASA workshop Professor Banks organized this summer on the role that the space station will play for the future of science in space.

Let me conclude this introduction with a brief historical note. The first LIDAR experiment used a ruby laser and was conducted not far from here more than twenty three years ago by Ron Collis of SRI International. He was, I believe, the one who coined the acronym LIDAR. The first LIDAR experiment that used a tunable laser to measure sodium in the upper atmosphere was conducted in 1969. Tunable lasers for remote sensing applications were reviewed more than a decade ago by Kildal and Byer. Dave Hinkley edited a book on remote sensing in 1976. A major milestone in NASA remote sensing was the completion of the Shuttle LIDAR working group study in 1979 under the leadership of Shelby Tilford and project scientist Ed Browell. That report detailed seven scientific objectives and twenty six separate scientific experiments that could be conducted from the orbiting shuttle. The report also pointed to the Nd:YAG and the CO_2 lasers as the probable transmitters of choice.

It is clear from the history of laser remote sensing, that the lack of adequate laser power and wavelength tuning range has held back the development of the field. I hope that this workshop identifies ways to overcome these longstanding limitations of the laser sources.

At the remote sensing conference held in Monterey, CA. in 1982, the backwater of underfunded remote sensing research was discovered by the Department of Defense. The increased Air Force, Army and Navy involvement in laser remote sensing, coupled with the President's Strategic Defense Initiative speech, has led to a Star Wars public image of lasers in space. For a number of us who have been involved with remote sensing research over the past decade and who have worked toward the application of remote sensing techniques to global measurements of scientific value, we hope that the NASA presence in remote sensing offers an alternate image of lasers in space, the Star Trek image of search and discovery.

Part I

Remote Sensing – An Overview

NASA Plans for Spaceborne Lidar: The Earth Observing System

R.J. Curran

Global Scale Atmospheric Processes Research Program,
National Aeronautics and Space Administration,
Washington, DC 20546, USA

Introduction

Observations of the Earth's atmosphere and surface from space platforms has made dramatic progress over the period of the last twenty-five years. A primary activity which occupied both scientists and engineers during this period was the design of passive imaging systems. The challenge of designing these systems was to make the delicate optics required to achieve reasonable spatial resolution such that it could survive the trauma of launch and still provide observations for a reasonable period of time. These instruments were designed to have widely differing spectral characteristics from the ultraviolet to the microwave. Engineering efforts were continually given to improving the lifetimes of both spacecraft systems and the observational instruments which, in turn, encouraged the development of standardized data processing hardware and software to display and analyze these data. These imaging systems have reached a high degree of sophistication in the Thematic Mapper instrument on Landsat and the High Resolution Infrared Sounder and the Microwave Sounding Unit on the operational NOAA spacecraft. Operational satellites are the logical application of techniques developed in the research and development process. The existence of such operational systems signals the maturity of the satellite Earth observations effort using passive sensors.

Scientific observations of the Earth using active sensors was initiated using the radar systems of Skylab in the early 1970's. Application of imaging radars and radar scatterometers continues to be an active research effort with instrumentation either on or planned for spaceflight.

The scientific impact of this new way of observing the Earth was demonstrated with the exciting geological discovery of ancient stream beds beneath the sands of the Saharan desert using the Shuttle Imaging Radar (see Elachi, et al., 1982 and McCauley, et al. 1982). Lidar observations of the Earth's atmosphere and surface offers the potential for equally exciting and scientifically useful results. With the urging of a number of advisory committees and the strong endorsement of the National Academy of Sciences, the NASA Earth Sciences and Applications Division is directing its activities to developing the next level in improvement of Earth sensing systems: the spaceborne lidar.

The logical development of any spaceborne system starts with laboratory or ground-based testing of the remote sensing concept. As is shown in Figure 1, the earliest ground-based systems developed under NASA sponsorship were in place and operating in the early 1970's. The second phase of the development utilized application of the lidar systems in an airborne laboratory environment. Several lidar systems have operated on the Galileo II (NASA CV-990) as well as other airborne platforms in which the experimenter can interact directly with the apparatus. The third phase utilizes the lidar in an autonomous, experimenter independent mode. The airborne platform used in this phase is the ER-2. This airplane flies at an altitude of 20 km for periods of 4 to 6 hours.

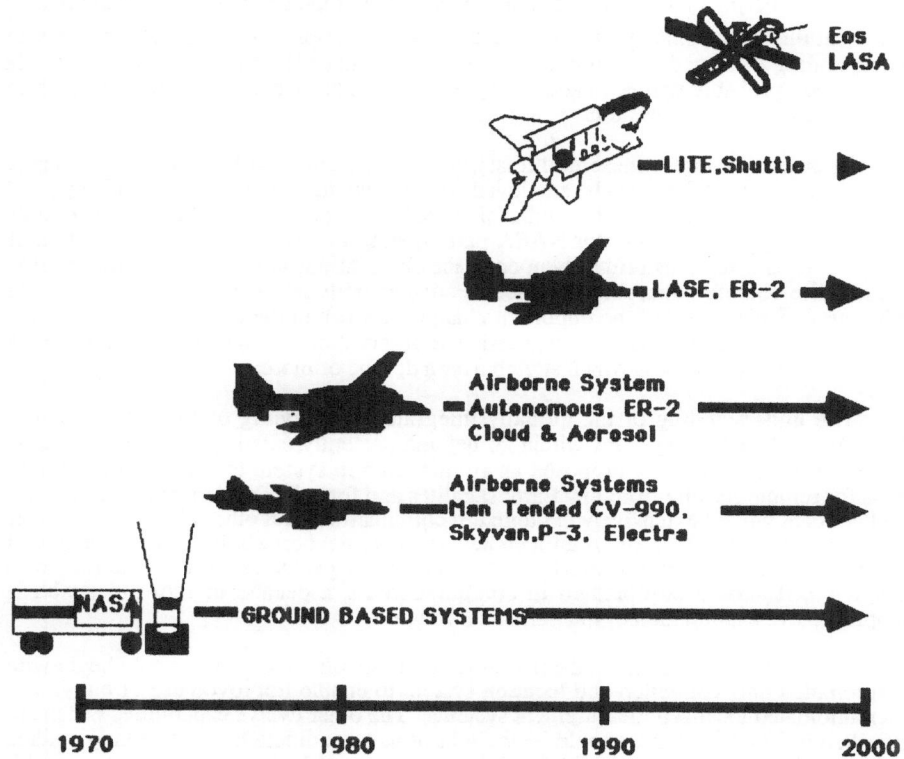

Figure 1. Chronology of NASA's program for the development of spaceborne lidar systems.

The first lidar system to be used on such a platform was an incoherent cloud and aerosol backscatter experiment flown in 1979 (see Spinhirne et al. 1982 and 1983). A differential absorption lidar experiment to measure atmospheric water vapor, the Lidar Atmospheric Sounder Experiment (LASE), is being developed for flight in 1987. It is anticipated that this facility will be used to test other lidar remote sensing techniques after its initial application.

A problem which is faced by any new sensor is the availability of a platform on which it may be placed. The Space Shuttle is a very useful platform to test both engineering concepts and remote sensing techniques. However, each of the earth sciences disciplines require periods of observations longer than that typical of Shuttle missions. For this reason, the Earth Observing System (Eos) with its polar orbiting platform planned for the mid-1990's would be ideal for the long-term observational requirement. It also will provide an ideal environment for multi-instrument observations of scientifically related phenomena. The general schedule going from ground-based experiments to the Eos platform is outlined in Figure 1.

<u>The Earth Observing System (Eos)</u>

Over the last decade or so a number of problems in Earth science have emerged which require a multidisciplinary approach. Examples include the

5

increase in atmospheric CO_2, the anticipated depletion of the ozone layer, El Nino related modifications to weather patterns, and acid precipitation. The key to progress in understanding these and other interdisciplinary issues in earth science during the decade of the 1990's probably will be in addressing those questions which concern the integrated functioning of Earth as a system.

In order to address these multidisciplinary problems confronting Earth science, observational capabilities must be employed which range in scale from detailed in situ and laboratory measurements to the global perspective offered by satellite based remote sensing. The potential future for NASA participation in contributing to the solution of these problems is to focus primarily upon space observations but with a clear recognition that satellite obtained data must be used in concert with data from more conventional techniques. Since many of the important changes at work in the earth system have time scales ranging from seasons to years, persistent observations of dynamic phenomena are needed to build data records which stretch over a decade or more.

The understanding of the globally integrated functioning of Earth will require observing and analysis systems which go beyond presently existing systems. An Earth Observing System (Eos) is proposed as an information system to meet many of these needs for remote sensing from low Earth satellites and for a data system to ensure that the resultant data will be extensively exploited in combination with other observables. Over the decade of the 1990's, a set of thirteen new measurement capabilities should be placed on orbit in a series of synergistically related groups or packages. When this full set is completed, it should be operated in combination for a decade or more to enable a comprehensive examination of the Earth as a system.

As is indicated in Table 1, the instrumentation set placed on orbit should begin with an automated data collection and location system to enable improved use of buoys and other automated localized measurement systems. The other twelve capabilities which are described in Tables 1 and 2 should be thought of as a candidate list of currently studied devices. It is recommended that flight of these instruments be in a sun-synchronous orbit with a 1:00 pm equator crossing time. It is assumed that one or more operational satellites may share platform space with Eos and, thus, will be in orbit in the same time period as Eos. These operational satellites will provide continuous, lower resolution soundings of the atmosphere that are not duplicated by the Eos. Conspicuous in Tables 1 and 2 are two lidar systems: the Lidar Atmospheric Sounder and Altimeter, and the Doppler Lidar Wind Sensor (see Earth Observing System, Science and Mission Requirements Working Group Report, 1984). A description of the objectives and the instrumental approaches for each of these instruments follows.

The Lidar Atmospheric Sounder and Altimeter (LASA)

As the title for this instrument implies, the scientific objective of the LASA package is to supply vertical profile information on several atmospheric parameters in addition to altimetry and retroranging to selected surface locations. Table 3 lists the general scientific objectives of the LASA instrument as well as the specific objectives to be addressed. The primary atmospheric parameter to be supplied by the LASA instrument will be the vertical and horizontal distribution of atmospheric water vapor. Water in its various forms is a key constituent in the Earth system. The transport of water vapor through large scale atmospheric motion is a segment of the hydrological cycle which is easily modified by a myriad of subtle processes in the Earth-atmosphere system. To better understand the processes involved in the hydrological cycle, accurate observations of water vapor with vertical resolution better than one kilometer are needed. Passive observations of atmospheric water vapor at either infrared or microwave wavelengths are limited in their vertical resolution. A lidar system on a polar orbiting satellite platform can provide high

Table 1. Earth Observing System Instruments – Data Collection and Imaging/Sounding Packages[*]

Instrument	Measurement	Spatial Resolution	Coverage
Automated Data Collection & Location System (ADCLS)	Data relay and Location	Location to 1 km for buoys and 1 m for ice sheet	global, twice daily

SISP – Surface Imaging & Sounding Package

Instrument	Measurement	Spatial Resolution	Coverage
Moderate Resolution Imaging Spectrometer (MODIS)	Atmos. and Surface Imaging 0.4 - 2.2 μm, 3 - 5 μm, and 8 - 14 μm	1 km^2 pixels	global, every 2 days
High Resolution Imaging Spectrometer (HIRIS)	Surface Imaging 0.4 - 2.2 μm	30 m × 30 m pixels	pointable to specific targets, 50 km swath
High Resolution Multifrequency Microwave Radiometer (HMMR)	1 - 94 GHz passive imaging	1 km at 36.5 GHz	global, every 2 days
Lidar Atmospheric Sounder and Altimeter (LASA)	Visible and near-ir range resolved backscatter	vert. res. 1 km, sfc. topography 3 m, retro-ranging 3 cm	global, soundings daily, topography 5 yr, retro-ranging every 6 months

[*] Modified from **Earth Observing System**, Science and Mission Requirements Working Group Report, 1984.

resolution observations of water vapor through range-resolved measurements of the differential absorption in and near a water vapor absorption feature. Water vapor lines at 0.723 μm and 0.94 μm are being considered (see Browell et al., 1979, and Cahen, et al. 1982) for this application, as well as lines near the 10.6 μm CO_2 laser emission line. Decision on which absorption line to use will await choice of the optimum laser transmitter for use on an extended space mission.

Two additional parameters of interest to the atmospheric sciences are temperature and pressure profiles. A complex technique proposed by Korb and his associates (see Korb and Weng, 1982 and 1983) is to use two very narrow spectral regions in the Oxygen A-band, at 0.76 μm, in conjunction with a nearby clear wavelength to obtain these two parameters. The technique is to use differential absorption at wavelengths which are independently sensitive to pressure and temperature. The wavelength stability and single mode requirements of the laser for this application are quite severe.

Table 2. Earth Observing System Instruments – Microwave and Atmospheric Monitor Packages[1]

Instrument	Measurement	Spatial Resolution	Coverage
SAM – Sensing with Active Microwaves			
Synthetic Aperature Radar (SAR)	L, C, and X-Band Radar Images	30 m x 30 m pixels	200 km swath daily coverage
Radar Altimeter	Topography of Oceans and Ice	10 cm vertical	global with repeating tracks every 10 days
Scatterometer	Sea sfc wind 1 m/s	minimum 1 sample every 50 km	global, every 2 days
APACM – Atmospheric Physical & Chemical Monitor			
Doppler Lidar Wind Sensor	Trop. winds to 1 m/s	1 km vert. 2°x2° lat. long	global, twice daily, Trop.
Upper Atmos. Wind Interfero.	Strat. winds to 5 m/s	3 km vert. 2°x2° lat. long	global, daily
Trop. Composition Monitor	Trace Constituents	dependent upon species	global, daily, Trop.
Upper Atmos. Composition Mon.	Trace Constituents	dependent upon species	global, daily, Strat. to 120 km
Energy and Particle Monitor	Solar Irradiance, Particles & Fields	total solar output	continuous

[1] Modified from Earth Observing System, Science and Mission Requirements Working Group Report, 1984.

An additional parameter of interest to the atmospheric sciences is the vertical distribution of Ozone in the Stratosphere and Troposphere. Again, the measurement technique would be differential absorption, using Ozone absorption lines in the Hartley-Huggins region of the spectrum (near 0.31 μm) for Stratospheric and Tropospheric observations. These Ozone observations could be studied in a synergistic approach with observations of other atmospheric parameters.

A by-product of each of the differential absorption measurements would be the vertical and horizontal distributions of atmospheric particulates as aerosols and clouds. These particulates serve as very useful tracers of the dynamical processes in the atmosphere as well as playing a significant role in the radiation energetics of the Earth's climate. Since the aerosol and cloud information comes from the elastic scattering of the

Table 3. Science Objectives for the Lidar Atmospheric Sounder and Altimeter on Eos

General	Specific
Contribute to a better understanding of the global hydrological cycle through observations of atmospheric water vapor	Water Vapor Column Content
	Water Vapor Profiles
Improve understanding of the role of aerosols and clouds in the earth/atmosphere system and their contribution to passive remote sensing from space	Total Aerosol Optical Thickness
	Tropospheric Aerosol Profiles
	Stratospheric Aerosol Profiles
	Boundary Layer Heights
	Cloud Top Heights
	Cloud Top Parameters
Improve understanding of the processes which influence ice and snow amount through altimetry	Precision Altimetry
Improve understanding of geodynamical processes through retroranging to globally distributed reflectors	Precision Retroranging
Provide globally distributed high resolution observations of atmospheric state parameters and Ozone to augment other earth system studies	Surface Pressure
	Pressure Profile
	Temperature Profile
	O_3 Column Content
	O_3 Profile

laser radiation, there is little requirement for wavelength stability and spectral purity of the laser transmitter.

An additional area of interest for the LASA instrument is the area of altimetry and retroranging. The requirements for altimetry are for vertical accuracies of the order of 10 cm. The major scientific goal of the altimetry measurements are to determine the ice/snow mass of the ice caps. These observations would be made every several years in order to monitor changes in the ice/snow mass and compare these changes with other climate parameters. Altimetry of other geographical locations could be made to augment our knowledge of the topography of these regions if this information is not available through other means. An additional result of the altimetry observations would be information on the vegetation canopy in selected regions due to range-resolved differences between surface height and canopy height. Retroranging from a satellite borne transmitter to specifically placed retroreflectors is a continuation of the highly successful scientific program using satellites in the inverted geometry. Mobile laser transmitters have been used near the San Andreas fault zone of California, to range to satellite-borne reflectors. These observations have been used to determine the relative motions of the Earths surface on either side of the fault line. The LASA retroranging experiment would be used to monitor a much larger number of fault zones globally, to better understand the overall dynamics of the tectonic plates. The requirements placed on the laser transmitters by the altimetry and retroranging science are for lasers producing pulses of duration less than one nanosecond.

The scientific objectives and observational approaches outlined above have been used in simulation studies to define the physical characteristics of the lidar systems. Several of the more important physical characteristics are tabulated together with their corresponding observational objective in Table 4. The above discussion implies the LASA instrument will be composed of several lidar systems. The spectral requirements of the atmospheric sounder portions of the LASA instrument, as indicated in Table 4, do not appear to be compatible with the pulse duration and pointing requirements of the altimetry and retroranging portions. The broad combination of requirements may be simply met with a combination of only three lidar systems identified as a single instrument package. In addition, the ability of lidar technology to meet the many requirements is different in each case. This variance in the ability of technology to match the instrument requirements' may force the LASA instrument package to change in time following its initial positioning in space. That is, the LASA instrument may necessarily have to be modified in space before it reaches its final configuration.

The Doppler Lidar Wind Sensor

Traditionally, the purpose of measuring atmospheric temperature and pressure using conventional means was to infer the horizontal density gradients. These density gradients together with the Earth's rotation determine the large-scale motions of the atmosphere. Because density gradient is calculated as a horizontal difference between two inferred densities, errors in inferring density at each of the two points are amplified into greater errors in inferring the atmospheric motions. For this reason, atmospheric scientists have sought direct observations of the large-scale motions of the atmosphere because of their higher quality. This is especially true in tropical regions where density gradients are more subtle. In recent years, CO_2 lasers operating in the 10.6 μm wavelength region, have been made stable enough to permit hetrodyne detection of the Doppler shift of the backscattered energy. The Doppler shift then can be used to directly infer the wind velocity. Such CO_2 systems have flown on aircraft platforms and successfully measured small scale atmospheric motions. These observations point to the future application of such coherent systems on space platforms. These satellite- borne systems may utilize the CO_2 laser technology which has been evolving for the past decade, or use the coherent solid state sources recently reported (see Kane et al., 1984).

Summary

Laser techniques offer a quantum jump in our ability to observe the Earth's atmosphere and surface from space. The increased spatial resolution and accuracy of these active systems will extend our understanding and predictability of processes which

Table 4. Lidar Instrument System Requirements as Based on LASA Science Objectives

Objective	Measurement Method	Wavelength	Energy	PRF[*]	Aperture
Retroranging	1 Wavelength[**]	0.5-1 μm	10 mJ	10	0.2 m
Altimetry	1 Wavelength	0.5-1 μm	50 mJ	20	0.5 m
Cloud Heights	1 Wavelength	0.5-1 μm	200 mJ	10	0.5 m
PBL Heights	1 Wavelength	0.5-1 μm	0.5-1 J	10	1.0 m
Stratospheric Aerosols	2 Wavelength	0.35 μm & 0.5 -1 μm	0.5-1 J	10	1.25 m
Cloud Parameters	1 Wavelength	0.5-1 μm	0.5-1 J	10	1.0 m
Troposheric Aerosols	2 Wavelength	0.5-1 μm	0.5-1 J	10	1.25 m
Column H_2O	2, DIAL[***]	0.727 μm	0.2-0.5 J	10	1.25 m
Column H_2O	CW DIAL	10-11 μm	10-20 W C	1.0 m
H_2O Profile	2, DIAL	0.727 μm or 0.93 μm	1.0 J	20	1.25 m
Surface Pressure	2, DIAL	0.76 μm	0.25 J	10	1.25
Pressure Profile	2, DIAL	0.76 μm	0.25 J	25	1.25 m
Temperature Profile	2, DIAL	0.77 μm	0.25 J	50	1.25 m
O_3 Total	2, DIAL	0.3 μm or 9.6 μm	------	--	-------
O_3 Profile	2, DIAL	0.3 μm or 9.6 μm	------	--	-------

[*] Pulse Repitition Frequency in Hz

[**] Retroranging requirement is for approximately 0.2 ns pulse duration

[***] Differential Absorption Lidar Technique

occur in the Earth system. The Earth Observing System offers a suitable platform from which to make these lidar observations in concert with other types of observations. The laser requirements of both the LASA and the Doppler Lidar Wind Profiler offer a challenge to those developing lasers to match these requirements.

References

Browell, E. V.,T. D. Wilkerson, and T.J. McIlrath, 1979: Water Vapor Differential Absorption Lidar Development and Evaluation, Appl. Opt.,18,3474.

Cahen, C., G. Megie, and P. Flamant,1982: Lidar Monitoring of the Water Vapor Cycle in the Troposphere, J. Appl. Meteor.,12, 1506-1515.

Earth Observing System, Science and Mission Requirements Working Group Report, Volume I, NASA Technical Memorandum 86129, 1984, pp.58.

Elachi, C. , W. E. Brown, J. B. Cimino, T. Dixon, D. L. Evans, J. P. Ford, R. S. Saunders, C. Breed, H. Masursky, J. F. McCauley, G. Schaber, L Dellwig, A. England, H. MacDonald, P. Martin-Kaye, F. Sabins, 1982: Shuttle Imaging Radar Experiment, Science, 218, pp. 996-1003.

Kane, T. J., B. Zhou, R. L. Byer, 1984: Potential for Coherent Doppler Wind Velocity Lidar Using Neodymium Lasers, Appl. Opt., 23, 2477-2481.

Korb, C. L. and C. Y. Weng, 1982: A Theoretical Study of a Two-Wavelength Lidar Technique for the Measurement of Atmospheric Temperature Profiles, J. Appl. Meteor.,12,1346-1355.

Korb,C. L. and C. Y. Weng, 1983: Differential Absorption Lidar Technique for Measurement of the Atmospheric Pressure Profile, Appl. Opt.,22, 3759

McCauley, J. F.,G. C. Shaber, C. S. Breed, M. J. Grolier, C. V. Haynes, B. Issawi, C. Elachi, R. Blom, 1982: Subsurface Valleys and Geoarcheology of Eastern Sahara Revealed by Shuttle Radar, Science, 218, 1004-1019.

Spinhirne, J. D., M. Z. Hansen, L. O. Caudill, 1982: Cloud Top Remote Sensing by Airborne Lidar, Appl. Opt., 22, 1564-1571.

Spinhirne, J. D., M. Z. Hansen, J. Simpson, 1983: The Structure and Phase of Cloud Tops as observed by Polarization Lidar, J. Clim. and Appl. Meteor.,22,1319-1331.

Solid State Laser Technology - A NASA Perspective

F. Allario

NASA Langley Research Center, Hampton, VA 23665, USA

With the advent of the Space Station Program within the National Aeronautics and Space Administration (NASA), significant opportunities to conduct lidar experiments from space will emerge in the decade of the 1990's to measure critical geophysical properties of the terrestrial atmosphere, land and oceans. The previous paper has discussed the concept of the Earth Observation System (EOS), with emphasis on those unique measurement requirements for lidar needed from a polar orbiting platform. Detailed scientific opportunities for a variety of remote sensor experiments can be found in a recently published NASA document.[1] In that document, it is apparent that many of the non-lidar experiments have a strong heritage in previous satellite and Shuttle flight experiments. In the case of lidar, the heritage lies in a long series of aircraft and ground-based experiments[2], in addition to a strong technology base which has been developed since the middle 60's[3]. Laser technology, just as microelectronics technology, is rapidly evolving with a concurrent need within NASA to maintain a strong technological base. For the last decade, NASA has sponsored research in gas lasers, including CO_2 and excimer lasers, dye lasers and to some degree in the ruby and Nd:YAG laser technology. For the next decade, NASA must develop a strong technology program in the emerging tunable solid state laser arena, in order to meet the scientific measurement needs of EOS. Solid state laser technology has the potential to meet EOS system requirements in efficiency, high reliability, long lifetime and modular design. An aggressive technology program is required to meet the projected scheduled requirements for the EOS program.

In order to begin definition of an aggressive technology program plan for NASA, this workshop was convened by the Office of Aeronautics and Space Technology (OAST). A major purpose of this workshop is to provide a process of information exchange where NASA scientists and technologists present their understanding of current needs for scientific measurements, present limitations of engineering design for conducting lidar experiments from space platforms, and provide a current perspective of laser technology that meets both science and engineering requirements from space platforms and space shuttle. The most important result from this workshop is for the science and technological community including university, industry and other government laboratories to recommend to NASA major elements of a tunable solid state laser technology program to meet flight needs of the next decade. This program plan must take into consideration the following factors: (1) the program must be synergistic with, and leverage other national interests; (2) the program must respond to NASA's unique scientific and engineering requirements; (3) the program must define those research and development efforts where NASA ultimately becomes a "smart buyer" of lidar flight systems; (4) the program must be realistic, trading off performance, cost and schedule. An important criterion in this program plan is the need for NASA to develop a program plan that rapidly infuses this new technology first into aircraft flight experiments, then into Shuttle experiments, and ultimately into the EOS platform. However, these exciting flight opportunities must be tempered with the reality that flight programs cannot be implemented without a strong technology program in materials research, materials characterization, and development of tunable solid state laser systems. The remaining portion of this paper will highlight some of the current research and system programs currently within NASA's planning cycle to meet this challenge.

TECHNOLOGY GOALS

The major goal for a NASA "Tunable Solid State Laser" program is to provide an assessment of an "all" solid state laser technology by 1988-89 to meet the emerging needs of the EOS program. With the current state of solid state technology, it appears that a major objective is to develop an "all" solid state pump laser in the green, using the emerging AlGaAs array technology, pumping a Nd:YAG/SLAB crystal, or glass. Some of the more important science wavelengths for EOS lidar experiments currently lie in the wavelength interval from 700-1000 nm. This wavelength interval is important to measure upper and lower tropospheric water vapor, aerosols, cloud heights, pressure, and temperature. Two solid state laser technologies which appear extremely attractive include alexandrite, and titanium doped sapphire (Ti: Al_2O_3). It is recognized that alexandrite is a relatively mature technology, and that relatively high energy systems have been recently developed. At this time, it must be recognized as a candidate laser technology for some EOS lidar experiments, with the long-range challenge to develop a reliable, long-lived flashlamp technology capable of at least 10^9 shots. Ti:Al_2O_3 is an extremely important technology since it can be pumped in the green, with an "all" solid state pump technology, and is tunable fromm 680 nm to at least 1000 nm. This combination of technology assets should provide the modularity and flexibility required for a series of EOS science investigations for a decade. A major challenge for the next several years is to bring Ti:Al_2O_3 to a technology readiness at the device level, for initiation of system definition studies and ultimately to system development for aircraft, shuttle and space platforms.

Another major science driver for EOS is a lidar system to measure tropospheric winds, to an accuracy of ± 1 m/s, with a vertical resolution of 1 km. Candidate technologies include long-lived CO_2 lasers at 10.6 microns, and Nd:YAG lasers at 1.06 microns. Tradeoff studies will be conducted over the next several years to define the optimal system from a global backscatter perspective, and from an instrument sensitivity point-of-view. Again, an "all" solid state laser in the green or in the near-infrared could serve as the basic technology to satisfy this important scientific need. Similarly, specific requirements are currently being generated to utilize this technology for laser altimetry.

14

In developing this technology plan, it is recognized that NASA needs to pay attention to the near-term needs discussed above, as well as the need to develop a long-range plan in materials growth and materials characterization of new promising solid state laser materials. In particular, NASA needs to play a major role in encouraging and conducting research in promising laser materials, and to develop a process of rapidly identifying laser materials for space science applications, and rapidly demonstrating the viability of these materials as laser transmitters for space science experiments. This should add to current national programs in tunable solid state laser development, by identifying NASA's unique requirements, and by growing NASA's engineers as "smart national buyers" for opportunities afforded by the EOS program.

Figure 1 depicts in pictorial form, the flavor which this program should take. Basically, in the early years an investment should be made in materials development and materials characterization, strongly coupled to research that quickly identifies promising laser materials. A strong linkage to development of selected solid state lasers must be developed with limited testing from aircraft and shuttle carriers as appropriate. Ultimately, the most promising technologies are to be carried onto space platforms, with the "payoffs" as identified on the chart. Over the coming year, NASA plans to assemble a program plan to efficiently implement the activities discussed in this truncated narrative. This program plan will be strongly influenced by the recommendations and strategies presented at this workshop.

SOLID STATE LASERS FOR SPACE APPLICATIONS

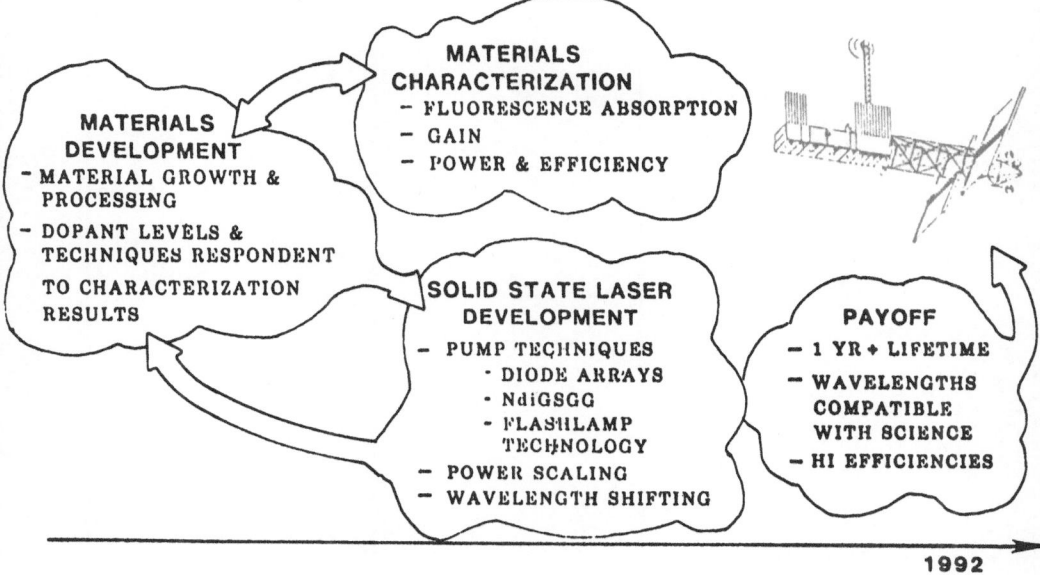

REFERENCES

1. "Earth Observing System: Science and Mission Requirements Group Report", Volume I [Parts 1 and 23]; NASA Technical Memorandum Number TM-86129; 1984.

2. "Airborne and Ground-Based Lidar Measurements of the El Chicon Stratospheric Aerosol from 90°N to 56°S." M.P. McCormick, et al; Geof Int., Volume 23-2, 1984. Pages 187-221

3. "Remote Sensing: A Proposed NASA Program for Space Missions in the Second Half of the Nineties' Decade"; By B. Rubin; Under NASA Contract Number NAS1-16000; 1 September 1984.

Coherent and Incoherent Lidar - an Overview

R.T. Menzies

Jet Propulsion Laboratory, California Institute of Technology, Pasadena, CA 91109, USA

The emphasis of this discussion will be on general concepts which are relevant to a comparison of coherent and incoherent LIDAR. Detailed systems comparisons for specific applications are beyond the scope of this paper. Concepts which are important in comparing and contrasting these two approaches to the detection of laser radiation include basic signal-to-noise (SNR) relationships, coherence diameter, speckle averaging, multiple pulse averaging, and the feasibility of single-frequency pulse generation from the laser transmitters. (Single frequency pulses are mandatory when using coherent LIDAR.) A great deal of work has been done in each of these areas, and the existing literature in these subject areas is vast. No attempt is made to include a comprehensive list of references; only a relatively small number are used as examples.

In Figure 1 are depicted schematic diagrams of the generic direct and heterodyne detection configurations, which are used in

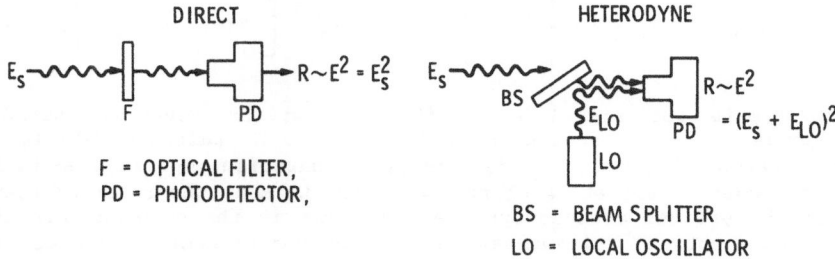

DIRECT

E_s ⟶ F ⟶ PD ⟶ $R \sim E^2 = E_s^2$

F = OPTICAL FILTER,
PD = PHOTODETECTOR,

HETERODYNE

E_s ⟶ BS, E_{LO}, LO, PD ⟶ $R \sim E^2 = (E_s + E_{LO})^2$

BS = BEAM SPLITTER
LO = LOCAL OSCILLATOR

(1) SPECTRAL RESOLUTION DETERMINED BY OPTICAL FILTER, TYPICALLY $0.1 - 10 \, \text{cm}^{-1}$

(2) DETECTION "GAIN" MAY EXIST, E.G., PHOTOEMISSIVE DYNODES, AVALANCHE EFFECT

(3) NOISE SOURCES: BACKGROUND RADIATION FLUCTUATIONS, THERMAL FLUCTUATIONS IN PD AND AMPLIFIER

(1) SPECTRAL RESOLUTION DETERMINED BY IF BANDPASS, TYPICALLY $10^{-5} - 10^{-1} \, \text{cm}^{-1}$

(2) DETECTION "GAIN" PROVIDED BY PHOTOMIXING WITH LOCAL OSCILLATOR

(3) THERMAL NOISE SOURCE EFFECTS CAN BE MINIMIZED WITH LARGE ENOUGH L.O. POWER. L.O. FLUCTUATIONS BECOME LIMITING NOISE

Figure 1. Differences between incoherent (direct) and coherent (heterodyne) detection.

incoherent and coherent LIDARs, respectively. (We'll treat homodyne detection, with zero offset between signal and local oscillator field frequencies, as a special case of heterodyne detection for the purposes of this discussion.) For efficient generation of the mixing term in the photodetector square-law response to the total electric field of the radiation, transverse phase coherence must exist across the detection surface, of course. The direct detection systems may limit the spectral bandpass using interferometers or dispersive elements in addition to filters; however the feasible spectral resolution limit is much larger than can be obtained with RF elements in the heterodyne receiver in any case. (A 1Å filter in the mid-visible corresponds to 4 cm^{-1} resolution.) For various reasons, coherent detection has been practiced almost exclusively in the infrared. The presence of high levels of detector thermal noise and thermal background radiation noise in the mid-infrared, and the lack of availability of photo-detectors which have high intrinsic gain in order to minimize the effect of electrical preamplifier noise (such as photomultipliers for visible wavelengths) have motivated groups to use heterodyne receivers for the detection of weak laser radar signals whenever suitable laser local oscillators were available. This combination of factors has resulted in the predominent use of coherent detection with CO_2 lasers in the 9-12 μm region. Advances in semiconductor diode and Nd:YAG laser technology have recently made possible the use of coherent LIDAR and laser communications receivers at shorter wavelengths.

The LIDAR equation for backscatter from a distributed medium is given as

$$P_{\lambda R} = P_T \cdot \beta_{\lambda R} \cdot \frac{ct_L}{2} \cdot \frac{A}{R^2} \cdot \eta_2 \cdot \exp\left[-2(\tau_{\lambda R}^e + \tau_{\lambda R}^i)\right], \quad (1)$$

where $P_{\lambda R}$ is the signal power incident on the receiver photodetector from the backscattering cell at range R, thickness ΔR; P_T is the transmittal power; t_L is the pulse duration; $\beta_{\lambda R}$ is the backscatter coefficient at wavelength λ, range R; A is the receiver area; η_2 is the optical efficiency; and the terms in the exponent are the integrated optical thicknesses due to the constituent i, and due to all other extinction sources excluding constituent i.

The receiver signal in Eqn. (1), as well as in other forms of the LIDAR eqn, which are appropriate to, e.g., fluorescence or topographic target returns, is proportional to the receiver (telescope) area; in this context a discussion of transverse coherence diameter is relevant. The refractive turbulence along an atmospheric path restricts the transverse coherence length and hence the maximum transmitter and receiver aperture sizes which are useful for coherent LIDAR. The general expression for transverse coherence length is [1,2]:

$$r_o = 0.058 \, \lambda^{6/5} \left[\int_0^L C_n^2(z) \, (z/L)^{5/3} dz\right]^{-3/5} \quad (2)$$

where integration is along the path of total length L, and $C_n^2(z)$ is the refractive index structure parameter, which may range from 10^{-13}. $m^{-2/3}$ to 10^{-15} $m^{-2/3}$ for paths near the surface, and which falls quickly with increasing height until it is generally two orders of magnitude smaller in the middle troposphere (5 km). For a coaxial lidar with correlation between the outgoing and return path refractive index variations, the correct expression is [1]:

$$r_a = r_0/(1 + \rho_{12})^{3/5}, \tag{3}$$

where ρ_{12} is a measure of the correlation between outgoing and incoming path fluctuations, taking values between 0 and 1. For a ground-based LIDAR system operating over a horizontal path length of a few km, at $\lambda = 10$ μm, r_a may range typically between 10-30 cm. For an Earth-orbiting coherent lidar operating at $\lambda = 1$ μm, r_a is $\approx 1m$ for paths extending down to the planetary boundry layer (altitude $\approx 1km$). Thus the tranverse coherence length may be an important consideration in the design of a coherent LIDAR system, whereas the incoherent LIDAR is not degraded by loss of transverse phase coherence due to refractive index inhomogeneities.

In order to estimate and compare the effectiveness of coherent and incoherent LIDAR systems for various applications, one must combine the LIDAR equation for received power with noise expressions and calculate signal-to-noise rations (SNR). We distinguish between what might be called the "traditional" $(SNR)_1 = \langle i(t) \rangle / [Var(i_0(t))]^{1/2}$, where $i(t)$ is the current when the signal and noise are present, and $i_0(t)$ is the current when only the noise is present, and a "better" SNR for LIDAR applications: $(SNR)_2 = (\langle i(t) \rangle - \langle i_0(t) \rangle) / [Var(i(t))]^{1/2}$, which is the difference between average values of the output current in the presence and in the absence of the signal, divided by the root-variance when both signal and noise are present [2,3]. The former definition, $(SNR)_1$, we call the carrier-to-noise ratio, CNR, and we call the latter the SNR. The CNR for incoherent backscatter LIDAR can be expressed as:

$$CNR = \frac{\eta \pi J \beta D^2 c e^{-2\alpha R}}{8(NEP)R^2} \tag{4}$$

where NEP (noise-equivalent power) is due to a combination of detector/preamp noise due to thermal fluctuations in carriers, and fluctuations in background radiation. For a 1 MHz bandwidth, typical NEP values at $\lambda = 10$ μm and $\lambda = 1$ μm are 3×10^{-11} W and 3×10^{-10} W, respectively. (In order to achieve this performance at 10 μm, the detector and preamp must be cooled to below 100K.) The CNR for coherent detection is

$$CNR = \frac{\pi \eta J \beta c \tau_p \, D^2 \, e^{-2\alpha R}}{8h\nu R^2} \tag{5}$$

In this case, the NEP $= h\nu/\tau_p$, i.e., quantum fluctuation noise

19

predominates in a well-designed heterodyne receiver [4]. For a matched-filter receiver, with 1 MHz bandwidth, values of $h\nu/\tau_p$ at $\lambda = 10$ μm and $\lambda = 1$ μm are 2×10^{-14} W and 2×10^{-13} W, respectively. Thus an advantage exists when coherent LIDAR is used in the ability to achieve higher values of CNR when the other parameters are similar. However, when calculating SNR, one must also account for fluctuations which are related to the signal itself, as was discussed earlier in defining $(SNR)_2$ as the LIDAR system signal-to-noise ratio. These signal fluctuations can be due to speckle effects or atmospheric refractive turbulence effects. The remainder of this discussion is limited to speckle fluctuation considerations. The resulting expression for SNR in the incoherent LIDAR case is:

$$(SNR)_D = \left[\frac{CNR}{(CNR/M_A M_T) + (CNR)^{-1}} \right]^{1/2} \tag{6}$$

where M_A = number of independent spatial modes which are detected and added to form the signal, and M_T = number of independent temporal samples which form the signal. The corresponding expression for SNR in the coherent LIDAR case is

$$(SNR)_H = \left[\frac{CNR/2}{1 + (CNR/2M_A M_T) + (1/2\ CNR)} \right]^{1/2} \tag{7}$$

The three terms in the denominator of Eqn. (6) are due to signal-mixing-with-noise fluctuations, signal-mixing-with-signal fluctuations, and noise-mixing-with-noise fluctuations [2,5]. When the CNR is high (large compared with unity) the signal fluctuations due to speckle, created in the scattering of the coherent laser radiation from a "rough" surface or aerosol, become important. The direct detection LIDAR systems are usually designed such that $M_A \gg 1$, i.e., receiver aperture averaging over many speckle lobes reduces these fluctuations. Aperture averaging in coherent detection may be accomplished using an array of photomixers, but in general the coherent detection LIDAR must rely more on temporal and pulse averaging in order to reduce the root-variance of the signal power.

For treatments of the combination of atmospheric turbulence and speckle effects on LIDAR, the reader is referred to References [2,5, 6] and other work cited in those references.

A number of studies of pulse averaging LIDAR returns indicate that various factors cause deviations from the standard reduction of the relative-root-variance of the signal by $N^{-1/2}$, where N is the number of returns being averaged, which is assumed to be valid when the pulse returns are uncorrelated [7]. The experimental evidence of Menyuk and Killinger indicate that returns from successive pulses are indeed somewhat correlated, i.e., the temporal auto-correlation coefficients,

$$\rho_j = \frac{1}{\sigma^2} \ <I(t)I(t + j\tau)> \tag{8}$$

where I(t) is the normalized LIDAR return at time t, τ is the pulse separation time, and j ranges from 1 to N, are not all equal to zero. This reduces the pulse averaging capability. Further studies of these effects under a variety of conditions are important.

References

1. S. F. Clifford and S. Wandzura, Applied Optics $\underline{20}$, 514 (1981).

2. J. H. Shapiro, B. A. Capron, and R. C. Harney, Applied Optics $\underline{20}$, 3292 (1981).

3. M. Elbanm and M. C. Teich, Optics Communications $\underline{27}$, 257 (1978).

4. R. T. Menzies, "Laser Heterodyne Detection Techniques: in Laser Monitoring of the Atmosphere, E. D. Hinkley, ed. (Springer Verlag, N.Y./Berlin, 1976).

5. R. M. Hardesty, "A Comparison of Heterodyne and Direct Detection CO_2 DIAL systems for Ground-Based Humidity Profiling", NDAA T. M. ERL SPL-64, October, 1980. (National Oceanic and Atmospheric Administration, Wave Propogation Laboratory, Boulder, CO.).

6. V.S.R. Gudimetla and J. F. Holmes, J. Optical Society of America $\underline{72}$, 1213 (1982).

7. D. K. Killinger, N. Menyuk, and W. E. DeFeo, Applied Optics $\underline{22}$, 682 (1983).

Remote Sensing from Space Platforms

E. V. Browell

NASA Langley Research Center, Hampton, VA 23665, USA

INTRODUCTION

Since its inception in the early 1960's, lidar has been extensively used for the measurement of atmospheric properties including molecular and aerosol backscattering, gas concentration profiles, wind velocities, and atmospheric waves. There are basically four main lidar techniques: elastic scattering, Raman scattering, resonance fluorescence, and differential absorption. Lidar measurement of elastic backscattering[1,2] from the atmosphere has been a prominent lidar activity that has focused on investigations of molecular, aerosol, and cloud properties. Lidar systems that detect returned signals at Raman shifted frequencies have been limited to measurements of gas species having high mixing ratios, such as water vapor, at ranges typically less than several kilometers[3]. Resonance flourescence has been an important lidar technique[4,5] to measure Na and K in the upper atmosphere, and it is the primary lidar technique for studying upper atmospheric waves[6]. The Differential Absorption Lidar (DIAL) technique can be used to measure concentration profiles of atmospheric gases at long ranges in the lower atmosphere[8].

During the past decade, NASA and the European space agencies have been conducting studies of spaceborne lidar systems. This paper will present the requirements for spaceborne lidar measurements of atmospheric species; the results of continuing studies related to the development of a spaceborne lidar; and the requirements for advanced lasers for use in spaceborne lidar systems.

SPACEBORNE LIDAR INVESTIGATIONS

Spaceborne lidar systems would enhance the ability to conduct atmospheric investigations over wide regions of latitude and longitude with unique emphasis on the troposphere. In addition, regions of the upper atmosphere that are difficult to study from the ground and with passive remote sensing techniques, like the ionosphere, would be easily accessible. To better define the role and capabilities of spaceborne lidar in conducting atmospheric investigations, several studies were conducted by NASA over the last ten years:

- Atmosphere, Magnetosphere, and Plasmas in Space (AMPS) Study (1975)
- Stanford Research Institute Study (NASA CR-132724, 1975)
- Lidar Proposals for Spacelab 1 (June 1976)

• Atmospheric Lidar Working Group Study (NASA Sp-433, 1979)
• General Electric Shuttle Lidar System Definition Study (NASA CR-3303, 1980)

Three major atmospheric research areas can be identified for the 1990's: atmospheric chemistry, climate, and weather prediction. The development of a spaceborne lidar system would contribute to an understanding of the processes governing the Earth's atmosphere and would also provide an evaluation of atmospheric susceptibility to manmade and natural perturbations. This paper discusses the measurement requirements for a spaceborne lidar system with emphasis on the major research area identified above.

ADVANCED AIRBORNE LIDAR MEASUREMENTS

While groundbased lidar systems have been used to measure atmospheric parameters from the troposphere to the mesosphere, the ability to perform lidar investigations from an aircraft platform adds a new dimension to atmospheric science research. The development and demonstration of lidar systems from aircraft serve to evaluate lidar techniques and system components prior to their application in a spaceborne lidar system. This section discusses the measurements made with the NASA Langley Research Center (LaRC) airborne Differential Absorption Lidar (DIAL) systems, and these measurements are related to future spaceborne lidar investigations.

The NASA LaRC airborne DIAL system[8] has been used over the past 4 years to measure O_3, H_2O, and aerosol profiles in the troposphere. The DIAL system operates in the ultraviolet for measurements of O_3 and in the near-infrared for measurements of H_2O. Along with the DIAL measurements in these wavelength regions, simulataneous aerosol profiles are obtained at 600 and 1064 nm. Ozone has been studied throughout the troposphere and into the lower stratosphere as part of the general objective to better understand the budget of O_3 in the troposphere. The airborne DIAL capability for the measurement of H_2O was developed to demonstrate the technique prior to applying it in space to global meteorology. An investigation of H_2O distributions over the Gulf Stream was the first application of this capability to a specific atmospheric investigation. The simultaneous measurement of aerosol has been used in all airborne DIAL investigations to define the boundary layer depth, cloud top heights, condensation levels, aerosol deliquescence levels, and stable layers in the free troposphere. With a multiple wavelength lidar aerosol measurment, there is additional information obtained about aerosol size distributions and possibly aerosol type. The results of airborne DIAL measurements of O_3, H_2O, and aerosols are presented in this paper, and future prospects for airborne and spaceborne DIAL measurements are discussed.

LASER REQUIREMENTS FOR SPACEBORNE LIDAR

Spaceborne lidar techniques require laser wavelengths that range from about 0.2 to 11 μm to make specific measurements from the thermosphere to the troposphere. Specific wavelength requirements for future spaceborne lidar systems are presented in this paper. In general, the UV wavelengths are more applicable for the upper atmosphere, and the IR wavelengths are more useful for the troposphere. The characteristics needed for spaceborne lasers are discussed in this paper. Most spaceborne lidar applications require tunable laser wavelengths with high average power. Solid state tunable lasers are attractive potential candidates for future spaceborne lidar missions.

REFERENCES

1. Fiocco R. T. H. and Grams G. 1966, Observations of the Aerosol Layer at 20 km by Optical Radar, J. Atmos. Sci. 21(3), 323.

2. McCormick M. P. 1975, The Use of Lidar for Atmospheric Measurements, Remote Sensing Energy Related Studies, Vezeroghu T. N. ed., Hemisphere Press, Washington, 113.

3. Cooney, J. A. 1971, Comparisons of Water Vapor Profiles Obtained by Radiosonde and Laser Backscatter, J. Appl. Meteo., 10, 301.

4. Bowman M. R. et al. 1969, Atmospheric Sodium Measured by a Tuned Laser Radar, Nature, 221, 456.

5. Megie G. et al. 1978, Simultaneous Nighttime Lidar Measurements of Atmospheric Sodium and Potassium, Planet Space Sci. 26, 27.

6. Rowlett J. R. et al. 1978, Lidar Observations of Wave-like Structures in the Atmospheric Sodium Layer, Geophys. Res. Lett. 5(8), 683.

7. Browell E. V. 1979, Ed., Shuttle Atmospheric Lidar Research Program-Final Report of Atmospheric Lidar Working Group, NASA, Spec Publ.-433.

8. Browell E. V. 1983, Airborne DIAL System and Measuremetns of Ozone and Aerosol Profiles, Appl. Opt. 22, 522.

Accommodation of a ND:YAG Lidar on a Polar Meteorological Satellite*

A. Rosenberg and D.B. Hogan

RCA Astro-Electronics, P.O. Box 800, Princeton, NJ 08540, USA

Spaceborne lidar sensors have the potential to measure atmospheric parameters - such as wind fields, humidity profiles, and vertical temperature structure - with the accuracy, spatial resolution, and coverage required by the meteorological community. Implementation of such operational lidars on today's civilian and military meteorological satellites using current laser technology is not feasible, principally because of the large size and high required-power of these systems.

Lidar technology has, however, matured to the point where it is of great importance to fly a lidar system now, on a current meteorological satellite, using available laser technology. Such a mission would provide vital information required for the construction and operation of the operational lidars. While the modest capabilities of the system would mean a limited operational role for the lidar, spaceborne lidar measurements would be demonstrated and important atmospheric parameters would be obtained. Furthermore, the experience gained by this mission would diminish some of the risks and uncertainties involved in the development of the operational lidar systems of the 1990's.

In a recent study for the Air Force, RCA Astro-Electronics has developed a conceptual design for a low-power, small-volume Nd:YAG lidar sensor which can be accommodated on a DMSP (Defense Meteorological Satellite Program) operational spacecraft. The characteristics of this instrument, designated the Spaceborne Lidar Sounder, are summarized in Table 1.

The entire instrument is packaged on an aluminum-honeycomb optical bench, and mounted on the earth-facing panel of the spacecraft, via a support structure that thermally isolates the sensor from the satellite. The optical bench also serves as a space radiator for thermal control. This mounting approach minimizes the impact of the lidar on the existing spacecraft configuration.

A preliminary analysis of the impact of the lidar operation on the spacecraft subsystems has been performed. In particular, the following areas have been addressed: power subsystem impact; field-of-view compatibility; launch weight, balance, and trajectory effects; command and data handling accommodation; thermal interface; and attitude control disturbances. Initial evaluations have shown that the lidar instrument can be integrated on a current DMSP operational

*This work was supported in part by the USAF Space Division, Los Angeles, under contract F04701-81-C-0061

TABLE 1   SUMMARY BASELINE LIDAR CHARACTERISTICS

	Parameter	Value
Transmitter	Laser Type	Nd:YAG Oscillator/Amplifiers and Second Harmonic Generator
	Energy Output	>1 Joule Pulse each at 0.53 μm and 1.06 μm
	Pulse Repetition Rate	10 to 20 Pulses per Minute (ppm)
	Output Aperture	0.1 meter
	Projected Life	0.5 to 0.9 x 10^7 Shots
Receiver	Telescope	0.5 meter f/6 Dall-Kirkham
	Primary	f/1.5 Lightweight
	Fields-of-View	0.05 mr Day/5.0-mr Night
	Detectors	0.53 μm and 1.06-μm cooled PMTs for Day and Night
System	Input Power	<120 watts
	Size	0.75 x 0.951 x 0.22 meters (less secondary mirror and sunshield)
	Weight	162 lbs.
	Operating Temperature	50°C
	Operation Life	12 to 20 months (at 10 ppm)

spacecraft (in addition to the currently planned payload), and operated for a period of 1-2 years, with minimal impact on the existing spacecraft design and configuration.

Solid State Bistatic Lidar for Ionospheric Species Mapping and Precise Locating Lidar for the Tethered Satellite System

T.D. Wilkerson

University of Maryland, College Park, MD 20742, USA

H.G. Horak, R.R. Karl, Jr., and D.J. McComas

Los Alamos National Laboratory, Los Alamos, NM 87545, USA

H.E. Spence

University of California at Los Angeles, Los Angeles, CA 90024, USA

Abstract

A new ionospheric measurement concept using a bistatic lidar is described for the Tethered Satellite System (TSS) and the Space Shuttle as suggested by McComas and Spence.[1] This technique will utilize low power lasers in the Shuttle bay. Lightweight photometers with small electrical consumption will be placed on the sub-satellite to be deployed on a tether towards the earth. Ranges will be from 20 to 100 kilometers from the Shuttle with a minimum sustainable altitude of approximately 130 km. The long duration of this mission offers unique opportunities for a global mapping of trace species in the 100-150 km altitude region. This technique offers sensitivities not currently available using ground-based or space-based instrumentation and provides opportunities for long-term monitoring at altitudes not accessible to non-tethered satellites. The two main lidar components are the following:

1. The "Laser Induced Fluorescence" (LIF) system employs a Shuttle-based tunable solid state laser to induce fluorescence along an atmospheric column adjacent to the tethered satellite.[2] Small (10 cm^2) photometers on the tethered satellite observe the laser-induced fluorescence at close range (~100 meters). The system´s sensitivity and detectable species concentrations are greatly improved by exclusion of the sunlit earth from the detector´s field of view. This permits the use of relatively low energy (5-50 mJ) lasers on board the Shuttle. In addition to the common approach using solid state lasers to pump tunable dye lasers, it is possible to use tunable solid state lasers. The Alexandrite laser accesses several atoms and molecules. Sodium can be excited with radiation from a two-wave mixing technique to combine two YAG lasers, giving high power in a narrow tunable output. This opens up the possibility of several new scientific investigations beginning with the second (tether down) TSS flight. These include measurements of (1) concentrations of atoms, ions, and molecules (2) basic chemical rate processes involving ionization, dissociation, reaction, and recombination, (3) global distribution and altitudes of neutral and ionic metallic layers (4) wind, tides, and atmospheric

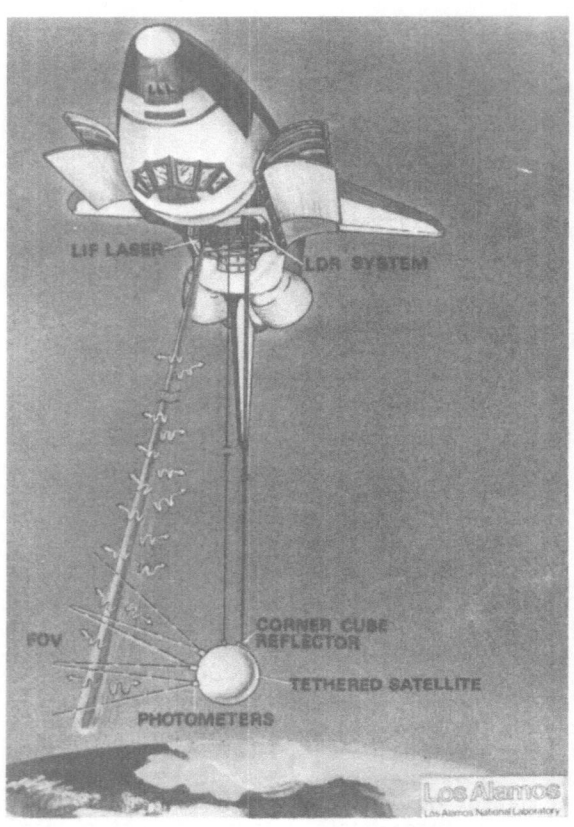

gravity waves (5) transport by diffusion, electromagnetic fields, and turbulence (6) response to external sources of light and particles.

2. The The second lidar system, called the "Laser Directional Ranger" (LDR) provides high accuracy, real-time range and direction angles of the tethered satellite in the Shuttle reference frame. The LDR relies on a Shuttle-based low-power, pulsed solid state laser, and a retroreflector mounted on the tethered satellite. It provides data on the tethered satellite location with accuracies of ˙2 meters horizontal and better than 1 meter vertical at 20 km range. These LDR data support the pointing of the LIF system so that the fluorescing ionospheric material is always located in the field of view of the tethered satellite´s photometers. Moreover, the data are more accurate than those supplied by other sensors, and are useful for detailed studies of TSS dynamics and the observation of anomalies in earth´s gravitational field. These instruments are useful for observations of opacity and scattering properties of released materials, including atoms, ions, molecules, and simulated interplanetary materials.

[1]D. J. McComas and H. E. Spence, "Space Shuttle Tethered Satellite System: A Unique Opportunity for Active Ionospheric Triangulation Experiments," Institute of Geophysics and Planetary Physics, UCLA publication number 2548, 9 May, 1984.

[2]Melvin I. Buchwald, Donald D. Cobb, Henry G. Horak, Robert R. Karl, Jr., David J. McComas, and John Zinn, Los Alamos; Paul J. Coleman, George L. Siscoe, Harlan E. Spence, UCLA; Marie-Lise Chanin, Service d'Aeronomie du CNRS; Chester S. Gardner, University of Illinois; Arieh Rosenberg, RCA Astroelectronics; and Thomas D. Wilkerson, University of Maryland. "Bistatic LIDAR for Ionospheric Species Mapping and for a Precise Locating LIDAR for the Tethered Satellite System," A Proposal Submitted to NASA, July 1984.

[3]C. R. Philbrick, J. L. Bufton, and C. S. Gardner, "A Solid State Tunable Laser Resonance Measurement of Atmospheric Sodium," Presented at NASA Workshop on Tunable Solid State Lasers for Remote Sensing, Oct. 1-3, 1984, Stanford University.

Part II

Solid State Lasers
for Remote Sensing

Nd:YAG and Ruby Based Lidar Systems for Remote Sensing of Atmospheric Aerosols

W.H. Fuller, Jr.

Mail Code 475, NASA Langley Research Center,
Hampton, VA 23665, USA

The investigation of atmospheric aerosols and their properties have become an increasingly active research area in recent years. This activity is a result of the realization of the importance of aerosols in such diverse areas as environment, climate, infrared backscattering, and cloud formation. In support of this research, lidar has proven itself to be a very valuable and versatile remote sensing technique. NASA-Langley has been involved with the development and application of lidar for atmospheric aerosol measurements since 1964. Over the years several lidars have been developed in the Aerosol Research Branch of Langley, leading to the current operational systems which include the ground based 48-Inch Mobile Lidar and a 14-inch airborne system. These lidar systems have been used primarily for stratospheric measurements, although recently, there has been increased activity in the tropospheric region. An 8-inch downlooking airborne lidar has also been developed, and is currently being used to support these measurements.

The 48-Inch Lidar has been in full operation since its development in 1972, and on a continuous basis since 1974. Its primary operational wavelength is 0.6943 µm (ruby laser) with additional capability at 1.06 µm (Nd:Glass laser). Measurements are also possible at the second harmonic generated wavelengths of 0.3472 µm and 0.5300 µm allowing four wavelength operation with the current lasers. The 48-Inch system utilizes photomultiplier tubes as detectors and uses a dual channel, computer based, high speed data acquisition system. The data system is unique with the capability of on line plotting of scattering ratio (total backscatter to molecular backscatter) profiles. The 48-Inch Lidar has monitored the stratosphere on a continuous basis since 1974 from the Langley site (37°N, 76°W) producing a unique time history data set as shown in figure 1. The aerosol peak backscatter ratio value (R-1) is plotted versus time; also indicated on the plot are significant volcanic eruptions during the 1974-1984 time period. The magnitude of the April 1982 eruption of the El Chichon volcano in Mexico compared to the other eruptions is quite evident in this figure. Using these lidar data, and supporting in situ size distribution and index of refraction deta, it is possible to establish a time history for the variation of the stratosphere aerosol mass loading and optical depth at a single point, not easily obtainable by any other technique.

Based on the experience gained with the 48-Inch Lidar, Langley has developed a 14-inch aperture dual wavelength, airborne lidar system. This system utilizes two laser transmitters, ruby (0.6943 µm) and Nd:Yag (1.064 µm) and uses photomultiplier tubes as detectors. The data system is similar to the previously discussed 48-Inch system,

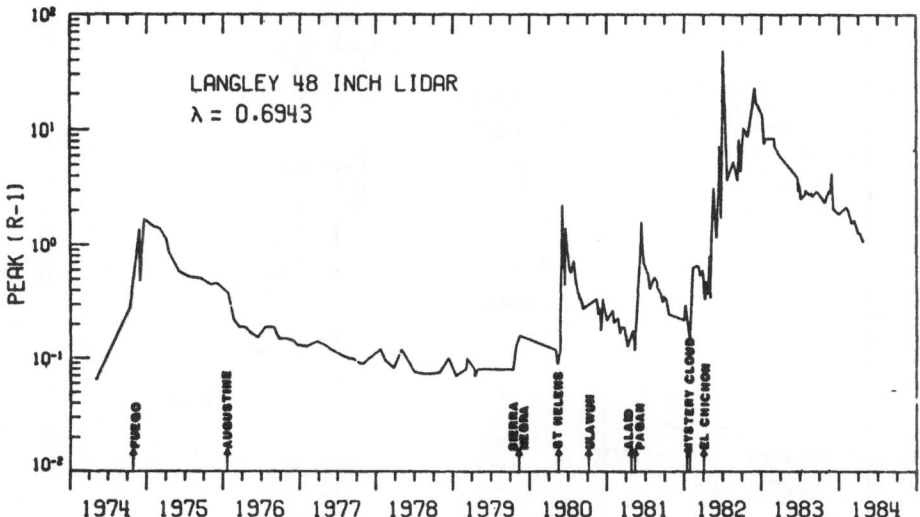

Fig. 1. Peak aerosol backscatter ratio versus time for ground based lidar data taken at 37°N, 76°W. Major volcanic eruptions are noted by arrows. (McCormick, M. P., et al., Geofisica Internacional, In press for October, 1984).

allowing hardware and software interchange. The airborne lidar data system also was designed with the capability of producing on-line scattering ratio plots, a valuable feature during airborne operations. The 14-inch airborne lidar was developed in 1978 to provide stratospheric aerosol data for correlative measurements in support of the SAGE (Stratospheric Aerosol and Gas Experiment) and SAM II (Stratospheric Aerosol Measurement) satellite instruments. Since that time, eleven satellite underflights have been conducted covering the latitude range 70°N to 15°S. In addition, five survey missions, to measure the latitudinal extent of the stratospheric material from the eruption of El Chichon, have been conducted ranging from 90°N to 56°S. An example of the lidar data obtained from one of the longer missions is shown in figure 2. Figure 2 is a plot of the integrated backscatter function versus latitude for the October-November 1982 mission. This value is derived from the airborne lidar scattering ratio profile data and is approximately related to total column loading or stratospheric optical depth. These data demonstrate the versatility of the lidar technique for detailed mapping of the aerosol spatial distribution.

In support of tropospheric measurement programs, an 8-inch downlooking airborne lidar was developed and utilizes Silicon diode or photomultiplier detectors depending on application and range. The system uses a Nd:Yag laser transmitter and the 14-inch lidar data system, but with additional intensity modulated plotting capability. The 8-inch downlooking system has been flown with the 14-inch uplooking lidar to obtain simultaneous stratospheric/tropospheric data. The 8-inch system has been applied to measurements of active volcanic plumes on three missions for the RAVE (Research on Atmospheric Volcanic Emissions) program and for recent Arctic haze measurements on the January, 1984 SAM II groundtruth mission.

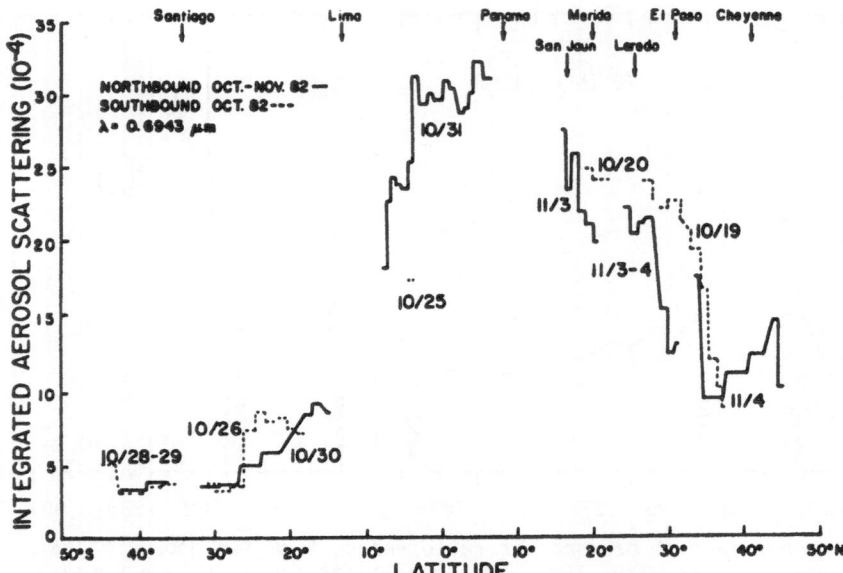

Fig. 2. Integrated aerosol backscattering function from the tropopause through the stratospheric layer versus latitude for southbound and northbound flights. (McCormick, M. P., and Swissler, T. J., Geophysical Research Letters, Vol. 10, No. 9, pp. 877-880, Sept. 1983.)

In summary, the lidar technique has been utilized most effectively in the study of stratospheric and tropospheric aerosols. The use of solid state lasers for these measurements has proven their reliability and operational efficiency. All flight missions have been highly successful with unique data sets that could not have been obtained with any other technique.

Remote Sensing with a Tunable Alexandrite Laser Transmitter

C.L. Korb, G.K. Schwemmer, M. Dombrowski, and R.H. Kagann

Goddard Laboratory for Atmospheres,
NASA/Goddard Space Flight Center, Greenbelt, MD 20771, USA

In this paper we describe a high resolution tunable Alexandrite laser system which we have used to make differential absorption lidar measurements of the atmospheric pressure profile. We also report on measurements of the spectral purity and line shape of the laser emission.

Our lidar system incorporates two Alexandrite lasers which were developed particularly for our application. They are continuously tunable from 725 to 790 nm and have a bandwidth of 0.02 cm^{-1} using a birefringent filter and two etalons. We have measured the short term frequency stability of the lasers to be better than 0.005 cm^{-1} and the Q-switched pulse length to be 100-130 nsec. One laser has a 100 mm x 5 mm diameter Alexandrite rod and a 150 mJ output energy. The tuning elements are electronically controlled and have a 3 cm^{-1} automatic spectral scanning capability. The second Alexandrite laser has a 75 mm x 5 mm rod and a 100 mJ output energy. Its tuning elements are manually controlled and it is typically used for the off-line measurement. Both lasers operate at a 10 Hz repetition rate and have a multi-mode spatial intensity distribution.

A 5 cm interferometer was used to make high resolution (0.002 cm^{-1}) measurements of the laser line shape by observing Fizeau fringes in transmission. The laser output consists of three axial modes with an overall width of 0.026 cm^{-1}. The individual modes have a width of 0.004 cm^{-1} and are nonuniformly spaced with a ratio of 1 to 1.3.

Spectral purity is an important parameter in lidar absorption measurements since spectrally impure out-of-band radiation (e.g., amplified spontaneous emission) may be weakly absorbed whereas the narrow-band component of the laser output may be strongly absorbed. For example, a 5% level of impurity can introduce a 100% error in a measurement with 95% absorption. We have shown that strong absorption lines can be used as a filter to reject the narrow band laser radiation. This allows the integrated out-of-band radiation of a pulsed laser to be easily measured. The spectral width of the rejection filter can be adjusted by varying the pressure of the absorbing gas. We have used this technique and found that the spectral impurity of our Alexandrite laser output is less than 0.01%.

Our ground-based lidar system has been used to make high accuracy measurements of the atmospheric pressure profile utilizing the integrated absorption in the wings of lines in the O_2 A band. We

use absorption troughs, regions of minimum absorption between two
strongly absorbing lines, for these measurements.[1] This technique
greatly desensitizes the measurements to the effects of laser
frequency instabilities.

The energy backscattered from the atmosphere is collected with a
45 cm telescope and detected with a photomultiplier tube. The
receiver field of view can be made as small as 0.25 mrad with an
adjustable field stop to provide rejection of daytime backgrounds.
In addition, narrow bandwidth (1 nm) interference filters provide
spectral background rejection. Separation of the on-line and off-
line laser pulses is accomplished by introducing a 100 µsec time
delay between the two laser pulses. A single detector channel is
then used to observe both wavelengths. The analog signals from the
photomultiplier are digitized to 10 bit accuracy at a 20 MHz sample
rate by a transient waveform recorder. This function, as well as the
system timing and computer interfaces to energy and spectral
monitors, is implemented using CAMAC (IEEE 583) standard modules. An
LSI-11/23 microprocessor controls system functions and monitors all
operator adjustable parameters.

An example of upward-viewing profiling measurements made with
our ground-based lidar system is shown in figure 1. The data were
taken at 20:30 EST on 3/25/83 with a 1 mrad field of view.
Interference filters were not used because of the low nighttime
background levels. The lidar system was set up to measure pressure
with the on-line laser tuned to the absorption trough at 13153.8 cm^{-1}
and with the reference laser tuned to a non-absorbing frequency near
13170.0 cm^{-1}. The lidar signal returns were sampled with a 50 nsec
range gate (7.5 m vertical integration) and averaged over 100
shots. The integrated absorption coefficient between the lidar and
each altitude was calculated from this data. The pressure profile

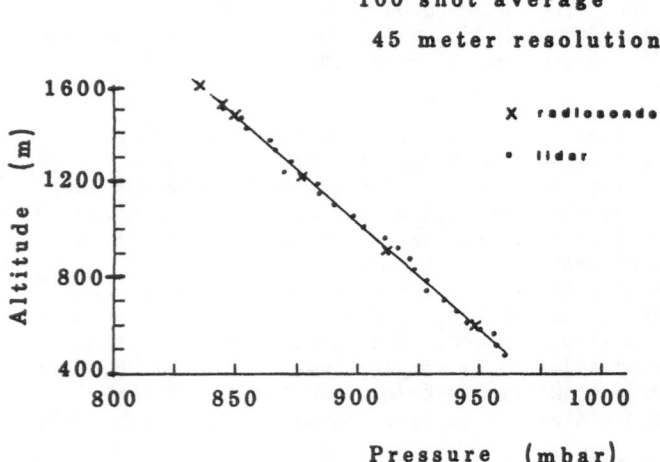

Figure 1. Lidar measurements of the atmospheric pressure profile
compared to radiosonde data.

was then determined using a simplified form of Eq. (9) in reference 1 which relates the measured integrated absorption coefficient to the difference in the squares of the pressures at the measurement altitude and laser altitude. Uncertainties in the oxygen line parameters were corrected for by a calibration fit of the measured data to ground truth. Figure 1 shows a comparison of the lidar measured pressure profile above Goddard Space Flight Center (lat. N 39.00^{0}, long. W 76.85^{0}) to radiosonde data taken 1.5 hours earlier at Dulles Airport (lat. N 38.98^{0}, long. W 77.46^{0}). The data were averaged to a 45 m vertical resolution in order to improve the signal to noise ratio. The average deviation of the lidar pressure data from the radiosonde data is 0.3%.

In addition to pressure measurements, our system will also be used to investigate temperature profile measurements using the high J lines of the oxygen A band.

References

1. Korb, C. Laurence, and Chi Y. Weng, <u>Applied Optics</u>, <u>Vol. 22</u>, No. 23, pp. 3759-3770, December 1, 1983.

Miniature Laser Diode Pumped Nd:YAG Lasers

T.J. Kane and R.L. Byer

Applied Physics Dept., Stanford University, Stanford, CA 94305, USA

ABSTRACT:
        Progress in high power laser diodes operating near 0.81 micron allow the construction of efficient, low power Nd:YAG lasers. We describe an end pumped Nd:YAG oscillator with 25% slope efficiency and 2 mW threshold. The demonstrated frequency stability of the diode pumped monolithic Nd:YAG laser is less than 10 kHz in 0.1 sec.

I. INTRODUCTION:

        The output of Gallium Aluminum Arsenide semiconductor diode lasers is very near the strongest pump band of the neodymium ion. Since semiconductor diode lasers can be extremely efficient (slope efficiency up to 70%) and since absorbed energy is converted into laser output energy with excellent efficiency in Nd:YAG (again up to 70%), diode pumping is an attractive means of obtaining efficient Nd:YAG laser operation. When small rods of Nd:YAG are pumped with the output of a single diode laser which is roughly mode-matched to the Nd:YAG laser spatial mode, threshold can be as low as 2 mW of pump power.

        Many researchers over the years[1] have demonstrated miniature diode pumped Nd:YAG lasers. Typically the diode pump sources operated at no more than a few milliwatts, and the Nd:YAG laser oscillated not far above threshold. The laser resonators have been simple, generally consisting of the Nd:YAG crystal and mirrors.

        Recent rapid advances in the power obtainable from diode lasers have led us to reconsider diode laser pumping of Nd:YAG oscillators. Continuous wave diode lasers with outputs of 80 mW and good spatial mode[2] or 2.3 W and poor spatial mode[3] have been demonstrated. These lasers can pump miniature Nd:YAG lasers far above threshold and achieve significant output power. It is possible to conceive of miniature Nd:YAG lasers of increased complexity, since even with the additional loss that results, threshold will be easily reached.

II. DIODE LASER PUMPED Nd:YAG:

        The key question to ask at this point is, in what way can a diode pumped Nd:YAG laser be more useful than the semiconductor diode laser that pumps it? Four attributes of a Nd:YAG laser that are impossible or difficult to achieve with diode lasers themselves are listed on the next page.

1. Amplifiability - Flashlamp pumped Nd:YAG amplifiers of very high gain exist.

2. Energy storage - The long upper state lifetime of Nd:YAG makes Q-switching possible.

3. Wavelength extendability - Internal doubling of miniature Nd:YAG lasers may be possible at very low power.

4. Frequency stability - The much higher cavity lifetime of a Nd:YAG laser leads to a much narrower Schawlow/Townes frequency stability limit than is possible from a diode laser, unless an external cavity is used.

In this talk we describe how the first three of these attributes might be realized with diode pumped Nd:YAG lasers, and then describe experiments in our laboratory which have demonstrated the excellent frequency stability of a diode pumped Nd:YAG laser.

1. Amplification by injection seeding. In injection seeding, a low power cw laser controls the oscillating mode of a much larger Q-switched laser. A very small cw power can create an advantage for a particular mode of the Q-switched laser, and thus establish single mode operation.[4] Strictly speaking, this is not the same as amplification. Nevertheless, it is a prime example of a case where a diode pumped Nd:YAG laser can be used with existing Nd:YAG technology in a way impossible with the diode laser itself. The two or three milliwatts that can be obtained when a miniature Nd:YAG rod is pumped with a commercially available 15 mW diode laser should be adequate to control the mode of a 0.5 Joule Q-switched Nd:YAG laser.

2. Energy storage and Q-switching. The neodymium ion can store energy over its upper state lifetime of 240 microseconds and then release it in the few cavity round trip times of a Q-switched pulse. This results in a large amplification of power. For a 5 millimeter miniature Nd:YAG laser, the round trip time is 60 picoseconds. Thus the pulse length will be in the range around 1 nanosecond. The ratio of storage integrating time to pulse length results in a power gain of 240,000, so a laser with a cw power of 10 mW could be converted into a pulsed laser of peak power 2.4 kW with a repetition rate in the low kilohertz range. If Q-switching were achieved using a saturable absorber film, then the overall laser would be very compact.

3. Wavelength conversion. The Q-switched laser described above would operate at a power level which could easily be doubled. Another technique for achieving visible output from a diode pumped Nd:YAG laser is by means of internal doubling. The round-trip losses in a miniature diode pumped Nd:YAG laser can be very small, substantially below 1%. Thus the internal circulating power of the laser can be over 100 times the cw output power. If the resonator contains a nonlinear material as well as the Nd:YAG crystal, doubling can take place. Circulating power near 1 watt may be doubled to the green with green output near 1 milliwatt. Compact, battery-powered lasers which are brightly visible may become possible.

4. Frequency stability.[5] Neodymium lasers are not generally considered for applications requiring good frequency stability. The index of refraction of Nd:YAG is a function of temperature, and thus fluctuations in the pumping or cooling process lead to changes in resonator optical path length and thus laser output frequency.

Flowing coolant is also a source of vibration which reduces stability. We have overcome these disadvantages through diode pumping of miniature Nd:YAG lasers. The diode pumping process can be made extremely stable by careful control of the diode temperature and current. The heat load is so small that conductive cooling is adequate. To eliminate another source of instability, we use the surfaces of the Nd:YAG rod itself as mirrors. The high rigidity of the Nd:YAG results in very high resistance to microphonics.

We fabricated two Nd:YAG rods of length 5 mm and diameter 2 mm. One end of each was coated for high reflection at 1.06 microns and the other was coated for 0.5% output coupling. One end was flat and the other had a curvature of 19 mm. The better of the two lasers reached threshold at 2.3 mW of diode output power and had a slope efficiency of 25%. Figure 1 is a plot of output at 1.06 microns as a function of diode output power.

We pumped both of the two miniature Nd:YAG lasers with a single diode laser and mixed the two outputs on a photodiode. The resulting beat frequency was observed using a frequency analyzer. The long term frequency drift was observed to be well under 1 MHz per minute. The frequency jitter over a period of 0.3 seconds was seen to be 10 kHz. Figure 2 is a frequency spectrogram of the beat signal obtained over the 0.3 second period.

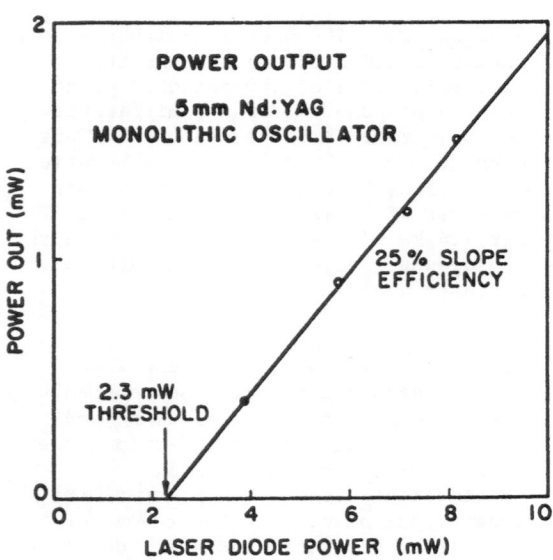

Fig. 1--Output power of Nd:YAG laser as function of output power of diode laser.

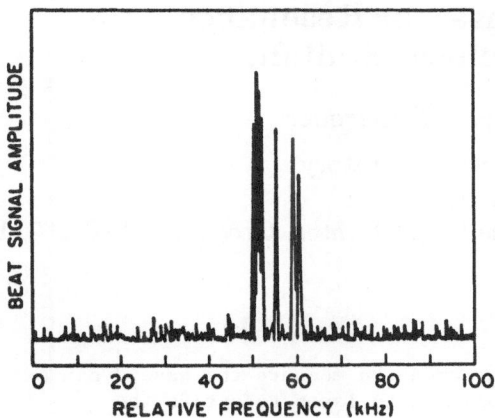

Fig. 2—Spectrogram of beat frequency between independent lasers,
taken over 0.3 seconds, shows 10 kHz jitter.

These results are from a system that employed no feedback
control at any level. With a closed loop system, extremely good
frequency stability should be possible, perhaps rivaling that of the
best helium neon lasers.

III. CONCLUSION:

Diode laser pumped Nd:YAG thresholds are measured to be 2 mW.
Since diode laser output powers near the 100 milliwatt level are
available, we project that good efficiency, reliability and lifetime will be obtained from a variety of diode pumped Nd:YAG laser
designs.

IV. REFERENCES:

1. For example, K. Washio, K. Iwamoto, K. Inoue, I. Hino,
S. Matsumoto and F. Saito, "Room Temperature cw Operation of an
Efficient Miniaturized Nd:YAG Laser End-Pumped by a Superluminescent
Diode", Appl. Phys. Letts. 29, 720 (1976).

2. H. Kawano, K. Endo, Y. Kuwamura and T. Furuse, "High
Power BCM Lasers with Low Astigmatism", paper 289, Annual Meeting of
Japanese Institute of Electronics and Electrical Communication, 1984.

3. W. Streifer, R.D. Burnham, T.L. Paoli and D.R. Scifries,
"Phased Array Diode Lasers", Laser Focus, June 1984.

4. Y.K. Park, G. Giuliani and R.L. Byer, "Single Axial Mode
Operation of a Q-switched Nd:YAG Oscillator by Injection Seeding",
IEEE J. Quant. Electr. QE-20, (1984).

5. B. Zhou, T.J. Kane, G.J. Dixon and R.L. Byer, "Efficient,
Frequency Stable Laser Diode Pumped Nd:YAG Laser", to be published
in Optics Letters.

A Solid State Tunable Laser for Resonance Measurements of Atmospheric Sodium

C.R. Philbrick[1], J.L. Bufton[2], and C.S. Gardner

[1]AFGL/LKB, Air Force Geophysics Laboratory,
Hanscom Air Force Base, MA 01731, USA
[2]Code 723, NASA Goddard Space Flight Center, Greenbelt, MD 20711, USA

The initial concept of an instrument to measure the wave dynamics in the upper mesosphere using a solid state laser to excite the resonance fluorescence line of sodium is discussed. The technique would make use of two Nd:YAG lasers to produce the sodium resonance line. The concept has been proposed for use in a space shuttle experiment.

INTRODUCTION

The most frequently used solid state laser is Nd:YAG which normally lases at 1064 nm, but another available transition for the same material is at 1319 nm. By mixing these two wavelengths in a nonlinear crystal, it is possible to generate a wavelength very near the sodium resonance lkne. In order to exactly tune the output to the sodium reaonance line, the wavelength of the 1064 nm laser can be tuned using an intracavity etalon. The feasibility of tuning the Nd:YAG laser to the wavelength required has been demonstrated in the laboratory (Marling, 1978). Even though no one has demonstrated the use of two Nd:YAG lasers to generate the sodium resonance line, all of the technical details needed to build such an instrument have been demonstrated. The idea of the sum frequency generation with Nd:YAG lasers, which makes this experiment possible, was first put forth by Aram Mooradian (private communication, 1983). The particular interest in generation of the sodium resonance line is to study the atmospheric waves and structure parameters using the meteoric sodium layer between 80 and 105 km as a tracer. A simple solid state laser for generation of the sodium resonance line could provide a major step toward a simple shuttle instrument for measurement of the horizontal scales of atmospheric waves.

LIDAR TRANSMITTER

In the previous developments of LIDAR instruments to measure Na, the laser transmitter has been a liquid dye (i.e. Rhodamin 6G), that is optically excited and tuned very precisely to the required wavelength. Dye lasers have the disadvantages of large size, system complexity, and a lack of ruggedized performance for the flight environment. The liquid dye and its fluid handling and cooling components are especially troublesome. In this effort, a fundamentally new and improved method is proposed for obtaining the required Na line radiation at 588.9 nm based on all solid state components. The primary laser source will be Nd:YAG lasers which have been commercially developed into compact rugged instruments suitable for being directly used in this flight

application. Several companies have rugged instruments for 28 volt dc
operation as off-the-shelf hardware. The method involves mixing the
1064 nm radiation with that from a second Nd:YAG operating at 1319 nm
in a nonlinear infrared crystal to directly produce 589 nm radiation
by sum frequency generation. Though not used as frequently, the
Nd:YAG laser can be operated at 1319 nm by inhibiting the normal
lasing at 1064 nm. In normal cw operation the output at 1319 nm is
about 30% of the output at 1064 nm (Marling, 1978). It will be
necessary to shift one of the wavelengths slightly to achieve the Na
resonance. This method is inherently quite simple and should provide
an effective all solid state source. The process of sum frequency
generation has been developed analytically by Armstrong et.al.,(1962).
The process is analogous to a second harmonic generation except that
two different wavelength beams of comparable intensity are input to
the crystal. By proper choice of crystal material, length, and angle-
of-incidence, efficient mixing under Type I phase matching can result.
The two Nd:YAG beams at 1064.1 nm and 1318.8 nm would result in an
output of 588.93 nm which is 0.06 nm from the required wavelength for
the sodium measurement. The work of Marling (1978) clearly
demonstrates that the 1064 nm line can be tuned over more than the
required range to produce the sodium wavelength required at 588.99 nm.
Figure 1 shows the schematic diagram for the transmitter.

TRANSMITTER SCHEMATIC

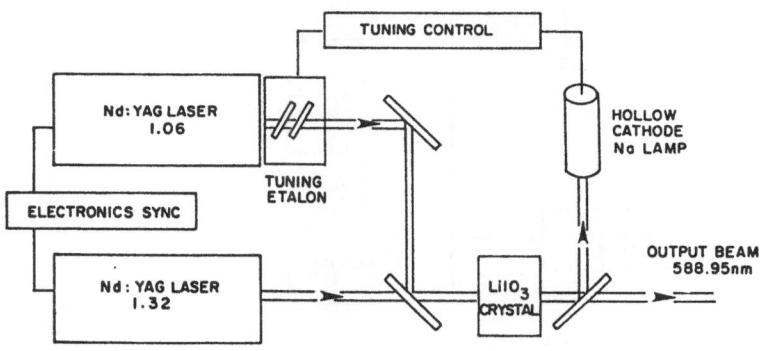

FIG. 1—Schematic diagram of the laser transmitter

CONCEPT FOR SHUTTLE EXPERIMENT

Plans have been developed to make use of a transmitter like that
described in the previous section to measure the sodium layer from the
space shuttle platform. The laser transmitter will require some
effort to demonstrate the measurement from the ground, initially.
The two lasers must be triggered together, the 1318 nm laser cavity
must be spoiled for the 1064 nm wavelength, the etalon tuned cavity
in the 1064 nm laser must be tested and the nonlinear crystal prepared.
None of the tasks are expected to cause any major difficulties. In
Figs. 2 through 5, the conceptual representation of the shuttle
experiment is presented. The plan is to make use of the GAS (Get-

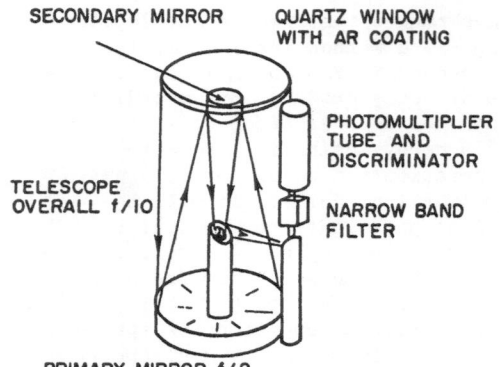

SECONDARY MIRROR   QUARTZ WINDOW
WITH AR COATING

FIG. 2--Diagram of receiver
configuration

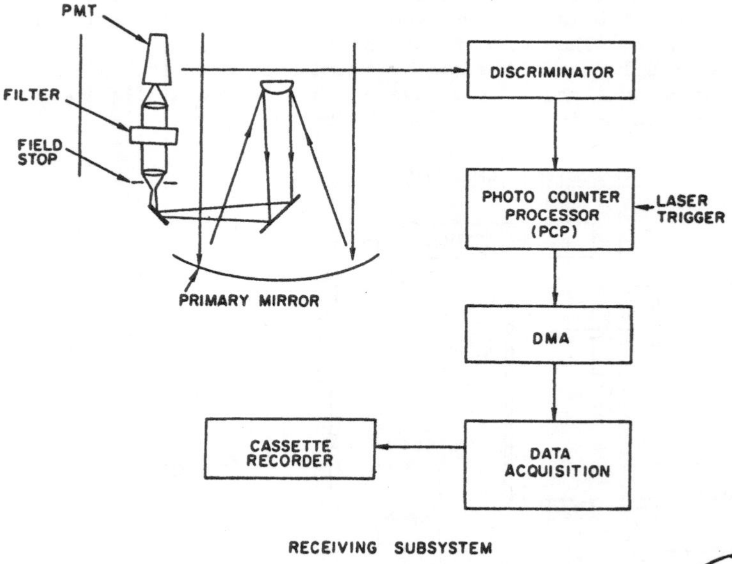

PHOTOMULTIPLIER
TUBE AND
DISCRIMINATOR

TELESCOPE
OVERALL f/10

NARROW BAND
FILTER

PRIMARY MIRROR f/2

PMT

FILTER

FIELD
STOP

PRIMARY MIRROR

DISCRIMINATOR

PHOTO COUNTER
PROCESSOR
(PCP) → LASER
TRIGGER

DMA

CASSETTE
RECORDER ← DATA
ACQUISITION

RECEIVING SUBSYSTEM

FIG.3--Components of the receiving sub-system

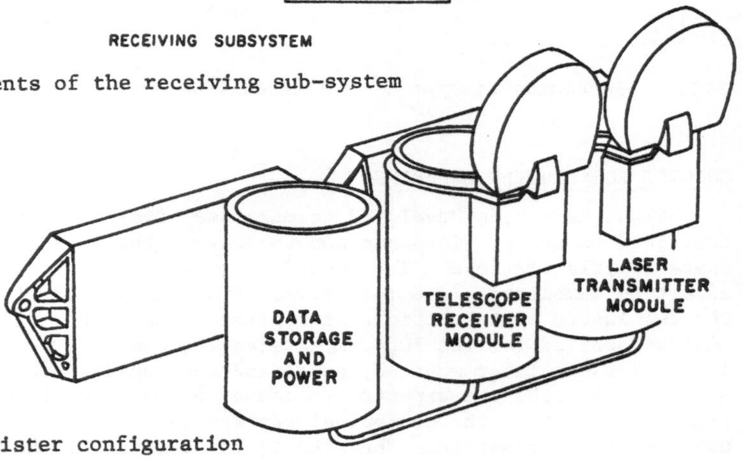

DATA
STORAGE
AND
POWER

TELESCOPE
RECEIVER
MODULE

LASER
TRANSMITTER
MODULE

FIG.4--GAS canister configuration

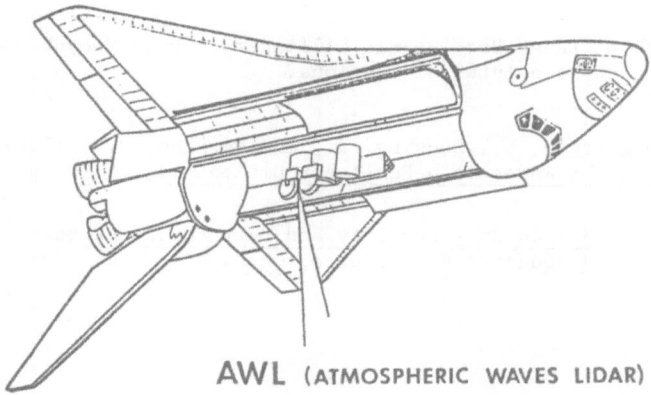

AWL (ATMOSPHERIC WAVES LIDAR)

FIG. 5--Shuttle experiment conceptual drawing

Away-Special) canisters of the shuttle program to make early
measurements of the horizontal scales of the waves in the sodium layer
and the global distribution of the sodium deposition. Table I shows
the parameters for the LIDAR system and Table II shows the expected
signal levels.

TABLE 1 - Lidar System Parameters

Primary Mirror	38 cm diameter, f/2
Effective Receiver Area	0.122 m^2
Field-of-View	10 mrad
Bandwidth	2 nm FWHM
Laser Energy	50-100 mj
Receiver Efficiency	0.06
Range Resolution	150 m
Integration Time	4 sec (20 shots)
Horizontal Resolution	32 km
Altitude Range	250-400 km
Effective Backscatter Cross-section	2×10^{-16} m^2 (10 pm linewidth)
	8×10^{-16} m^2 (1 pm linewidth)

Table 2 - Expected Signal Levels

Z (km)	E_T(mj)	Linewidth (pm)	N^*_{tot}
250	50	10	14-56
250	100	10	28-112
250	50	1	56-225
250	100	1	112-450
400	50	10	3.7-15
400	100	10	7.5-30
400	50	1	15-60
400	100	1	30-120

*Column Abundance $= 2$ to 8×10^{13} m^{-2}

45

ACKNOWLEDGEMENTS

The authors wish to thank Drs. Dennis Killinger, Norman Menyuk and Dwight Sipler for their helpful discussions.

REFERENCES

Armstrong, JA., Bloembergen, N., Ducuing, J., and Pershan, P.S. (1962) Interactions Between Light Waves in a Nonlinear Dielectric, Phys.Rev. 127, 1918

Marling, J., (1978), 1.05-1.44 μm Tunability and Performance of the cw Nd^{3+}:YAG Laser, IEEE J. Quantum Electronics, QE-14, 56

Part III

Advances in Laser Concepts -
Diode Pumped Nd:YAG

Laser Diode Arrays for Pumping Nd:YAG

M. Ettenberg

RCA Laboratories, Princeton, NJ 08540, USA

The practical use of solid state lasers in space is presently precluded by the low efficiency and limited lifetime of the flash-lamp pumps. Laser diodes are an ideal solid state long-lived replacement for this flashlamp tube and while it is believed that in the long term laser diodes will also replace the solid state rod and be used as the primary laser source, the interim solution is to use the laser diodes in a two dimensional array to pump a Nd:YAG slab to create high peak power pulses of coherent radiation. The slab acts as a coherent combiner and a pulse width compressor for the more efficient but relatively lower radiance laser diode arrays.

The requirements on the laser diode sources are relaxed as compared to the more critical applications such as optical recording and fiber optic communications and we can rely on the older and better established technology of oxide-stripe gain guided devices. To this end we have made 2 mm long arrays with an output power of 8 W (1 μs long pulses at 300 Hz) with emission at a peak absorption of Nd:YAG, 8085 Å. Light output power to total electrical input efficiency of the conversion from one facet of these arrays (the rear facet was coated with a high reflectivity dichroic reflector [1]) was 33%. Such arrays were made using our oxide stripe processing on AlGaAs double heterojunction material with an active region of 800 Å. Similar oxide stripe devices using the same growth and processing have demonstrated operating life in excess of 10^5 hours at room temperature [2]. This operating life applied to our arrays yields a pump which should survive for more than 10^{12} shots from the Nd:YAG laser.

The linear arrays are mounted on BeO submounts and incorporated into a copper block, producing a two dimensional array 0.4 x 1 cm containing 112-2 mm long linear arrays producing an expected power density well in excess of 1500 W/cm^2.

While the production of such two dimensional arrays is a fairly straightforward packaging task based on existing device technology, it will involve making enough of these devices so that the learning process can work to provide high yield and reproducibility to reduce cost. The still open issue has to do with the spectral output of the laser diode array and the resulting absorption efficiency in the Nd:YAG. The absorption lines in Nd:YAG are fairly narrow; while the spectral output from a single linear array can be as narrow as a single spectral line (<0.1 Å) it is more typically 20 Å (a few spectral lines spaced at about 4 Å). Even if we select linear arrays emitting only within such a narrow spectral bandwidth to populate the array we will have some variations in thermal and electrical resistance and might expect a 3-6°C variation in active region temperature across the array leading to a 10 to 20 Å variation. Another contributor to spectral variation is spectral shifts with aging [3] which can account for another 20-30 Å shift. These variations in spectral output will have to be studied in real arrays to see if they are additive and to determine if such spectral shifts will have to be accounted for in the diode laser pump/solid state crystal cavity design.

REFERENCES

[1] M. Ettenberg, "A New Dielectric Facet Reflector for Semi-conductor Lasers," Appl. Phys. Lett., 32, 724 (1978).

[2] M. Ettenberg, H. Kressel and I. Ladany, "Long-Term Lifetests of C.W. (AlGa)As Laser Diodes at Room Temperature," Electron. Lett., 14, 815 (1978).

[3] D. Botez, J. C. Connolly, M. Ettenberg, D. B. Gilbert and J. J. Hughes, "The Reliability of Constricted Double-Hetero-junction AlGaAs Diode Lasers," Appl. Phys. Lett., 43, 137 (1983).

Progress in Diode Array Development

D. Scifres, P. Cross, and H. Kung

Spectra Diode Labs., 3333 N. First Street, San Jose, CA 95134, USA

The goal of multiple semiconductor lasers being integrated onto a single semiconductor chip has recently been realized both in the research lab[1] and in the commercial arena.[2] These laser arrays, because of their high optical power density, high efficiency, small size, long life and emission wavelength (which can be made to match the absorption band of Nd:YAG lasers) may offer an attractive alternative to conventional lamp pumped solid state lasers.

The assembly of multiple diodes onto a single chip on a commercial basis is a direct result of the reproduceability, uniformity and reliability of devices produced by the metalorganic chemical vapor deposition growth technique which is used for wafer production. Since this technique is well suited for high volume mass production it is further believed that low cost arrays of even larger size and higher output powers can be achieved in the future.

To date, multiple quantum well lasers with 10 emitters on 10 μm centers (see Fig. 1) have achieved output powers of up to 600 mW/facet under cw room temperature operation. Continuous wave power versus current curves for such devices at heat sink

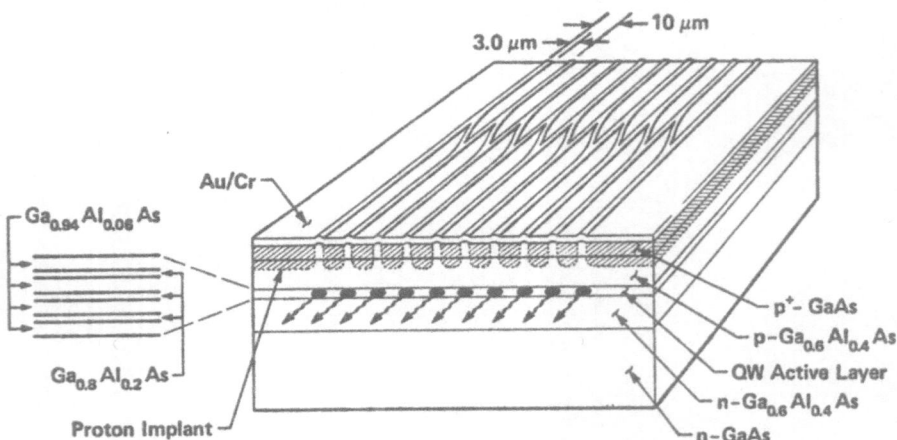

Fig. 1 Schematic diagram of a 10 emitter multiple quantum well phase locked laser diode array grown by MOCVD.

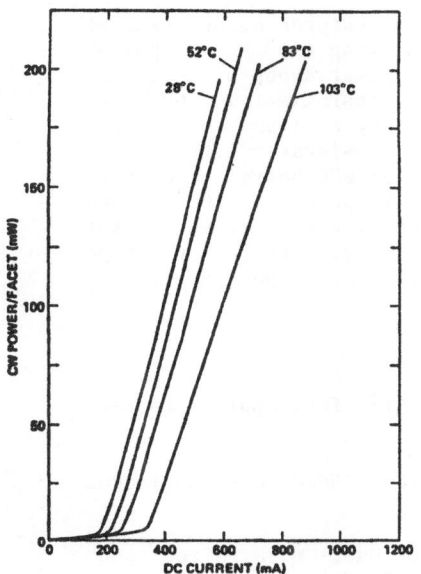

Fig. 2   CW power versus current
          for a 10-emitter laser
          diode array operating
          at various heat sink
          temperatures

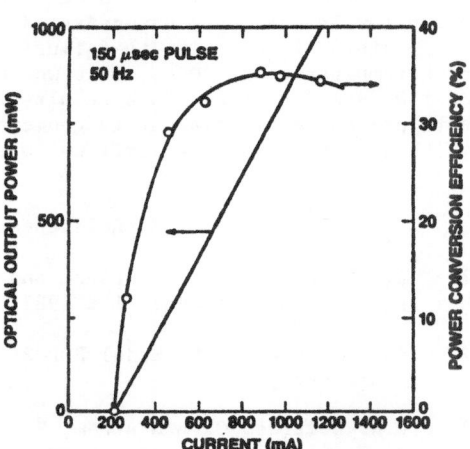

Fig. 3   Experimentally measured
          power output and overall
          efficiency as a function
          of drive current for a
          commercial SDL 10-stripe
          laser diode array. A
          peak conversion effi-
          ciency of greater than
          35% was demonstrated.

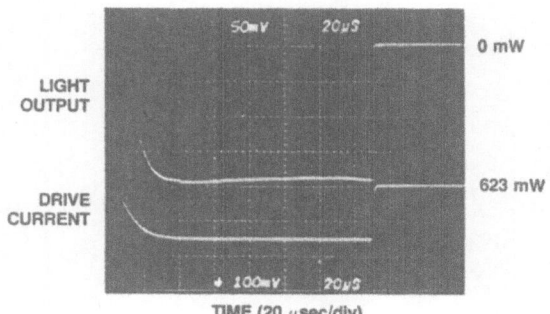

Fig. 4   Measured light output and drive current as a function of time
          for a 10-stripe, commercial SDL device

temperatures in excess of 100°C are shown in Fig. 2.   Devices
operating under pulsed conditions (150 μsec pulse width, 50 Hz)
have been operated to output power levels of 1 watt/facet with a
differential   quantum efficiency of approximately 70% and a power
conversion efficiency of up to 36% (see Fig. 3).   In contrast to

51

conventional lasers, quantum well laser structures do not exhibit nearly as large a power droop during the long (150 μsec) pulse. This is illustrated in Fig. 4 where a power droop of ≈ 5% is measured at an output power level of approximately 600 mw. Finally commercial laser diodes operating at 100mw cw output power levels have been demonstrated, via accelerated lifetesting, to exhibit a mean time to failure of ~ 31,000 hours at a heat sink temperature of 30°C. This would correspond to ~ 10^{12} pulses of 150 μsec duration. Such results foretell a widespread use for both present commercial array geometries as well as larger arrays which have been shown to emit up to 2.6W of cw output power.[3]

References

1. D.R. Scifres, R.D. Burnham and W. Streifer, Applied Physics Letters, No. 41, Vol. 118 1982.

2. Spectra Diode Labs Model SDL-2410 is a 100mW cw phase locked diode array.

3. D.R. Scifres, C. Lindstrom, R.D. Burnham, W. Streifer and T. Paoli, Electronics Letters, No. 19, Vol. 169, 1983.

Part IV

Review of Tunable Solid State Laser Systems

Review of the Tunable Alexandrite Laser Performance

J.C. Walling

Allied Corporation, 7 Powderhorn Dr., Mt. Bethel, NJ 07060, USA

Summary

The growth of alexandrite crystals by Czochralski methods was begun at Allied Corporation in the early 70's by Carl Cline and Robert Morris. The crystals produced were adequate for laser rod fabrication and preliminary laser work was performed. In 1977, Hans Jenssen and John Walling discovered tunable, vibronic laser emission from alexandrite. This discovery led Allied to develop the material for commercial and military markets. Since then, other similar solid state lasers having the same basic (vibronic) mode of operation, have been discovered increasing the scope of material properties available to the laser designer. Alexandrite has a unique compliment of intrinsic properties that provide high performance, comparable to Nd:YAG, ruby and Nd:glass, today's dominant fixed frequency solid state soruces. Tunability, and extended versatility, based on a temperature dependent emission cross section, should also be emphasized. The performance obtained from alexandrite indicates a comparatively advanced level of development among lasers of this class of operation.

Alexandrite is a Cr^{3+} doped chrysoberyl (chemical formula $BeAl_2O_4$). It is isomorphic with olivine and has the orthorhombic space group D_{2h}^{16}. Cr^{3+} substitutes on two sites normally occupied by Al^{+3}. The more occupied site, with 78% of the total Cr^{3+}, possesses mirror symmetry and is dominant in the laser action. The second site, possesses inversion symmetry, and plays a minor role in laser action. Physically, alexandrite is a hard, gem like crystal with high thermal conductivity (2/3 that of sapphire) and therefore it possesses excellent qualities as a laser host.

While alexandrite is basically a 4-level laser having a vibrationally excited ground state as a terminal level, it is unique in having a fifth storage level, about 800 cm^{-1} below the initial lasing level, that temporarily traps excitation and extends the storage time (at the expense of laser gain). This property imbues alexandrite with a strongly temperature dependent emission cross section and, hence, one that may be temperature taylored for each application.

Alexandrite is effective both as a pulsed and cw laser. It has good storage capability in Q-switched operation. In this mode of operation, it is capable of storing 1 to $3J/cm^3$ of extractable energy. Thus, alexandrite provides a compact high-energy laser source. Alexandrite multimode oscillators with rods 6.35 mm in diameter and 11 cm in length typically yield > 500 mJ, 30 ns ~5X diffraction limited pulses at 20 Hz repetition rate. Alternatively, these rods produce 2 J, 40 ns, 25X diffraction limited pulses at the same repetition rate. Alexandrite lasers have been operated single mode producing pulse energy in the few mJ range. Experiments are underway to improve performance in single mode operation.

For applications requiring higher brightness, alexandrite oscillator-amplifiers may be used. Alexandrite single pass amplifiers typically produce amplification factors of 2 to 5 for large signal inputs. Alexandrite oscillator-amplifier systems that utilize 0.64 cm diameter rods provide about 100 MW of peak power in a 5X diffraction limited beam, adequate for reasonably efficient frequency conversion by raman or nonlinear mixing processes. Frequency conversion studies with the alexandrite oscillator amplifiers have been very encouraging; 200 mJ tunable pulses at the second harmonic frequency of 380 nm have been generated in KDP at a repetition rate of 10 Hz. In raman shifting experiments, using the oscillator only, over 30% energy conversion into the first Stokes has been achieved using a single pass H_2 gas cell.

Fig. 1 shows the performance of an alexandrite single pass amplifier at constant pump power as a function of input energy. The curve in Fig. 1 is the predicted performance. The slight fall off at higher injected energies shows the effect of saturation. Even at 3.5 J output, with 2.5 J derived from the 4.5 mm apertured, 11 cm long, ampli-

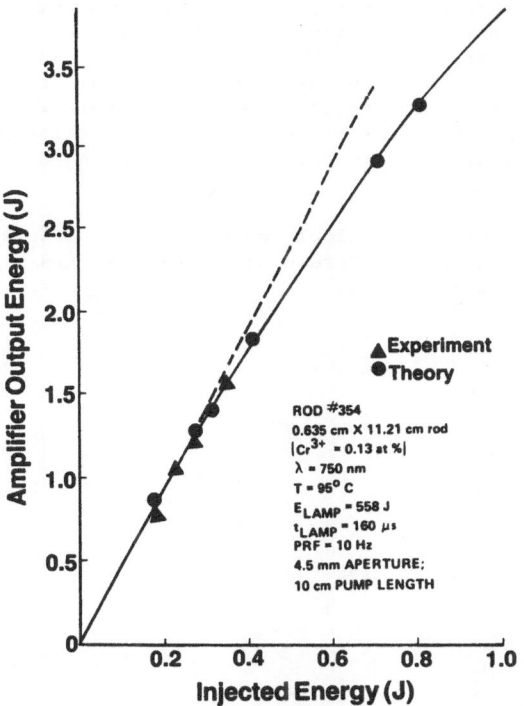

▲ Experiment
● Theory

ROD #354
0.635 cm X 11.21 cm rod
|Cr^{3+} = 0.13 at %|
λ = 750 nm
T = 95° C
E_{LAMP} = 558 J
t_{LAMP} = 160 μs
PRF = 10 Hz
4.5 mm APERTURE;
10 cm PUMP LENGTH

Fig. 1. Alexandrite amplifier performance: comparison with theory.

fier rod, only a small saturation effect is both predicted and observed. This underscores the large energy storage capability of alexandrite.

Studies with mode-locked, pulsed alexandrite lasers, at Allied, have used a linear colliding pulse resonator that features two counter-propagating pulses which overlap in the optical center of the resonator. Saturable absorbers are placed at this position to generate passively mode-locked, transform-limited, 38 ps pulses. In general, such pulses are tunable over a broad band (with 50 ps pulses, tunability from 735 to 768 nm has been achieved). Combined Q-switched and mode-locked operation of alexandrite has resulted in almost 2 mJ per individual pulse extracted from the oscillator.

CW alexandrite lasers have been both arc lamp and ion laser pumped. Highest power performance to date, 60 W, has been obtained using Hg lamp pumping. Despite the low emission cross section, experience has shown that alexandrite is efficacious in the cw mode where it competes well in general performance with Nd:YAG, but has the additional highly advantageous property of tunability.

For LIDAR applications, the alexandrite's tunability and spactral purity are well suited for DIAL measurements of pressure, temperature, and water vapor content of the atmosphere. Raman frequency conversion of the alexandrite fundamental can conveniently address the 940 nm water band conveniently. Extended tunable coverage of the UV through IR can be obtained from alexandrite by a number of frequency conversion techniques. Alexandrite's high spectral purity is a consequence of the low cross section and is key to sensitive DIAL measurements. In this respect alexandrite is substantially superior to dye lasers. From the standpoint of system size and complexity, a compact, line narrowed alexandrite oscillator can replace much larger and more complex alternatives, e.g., a frequency doubled, amplified Nd:YAG laser pumping a line narrowed dye laser.

Acknowledgments: The author wishes to achnolowledge his many coworkers who have contributed the results described herein and whose work has been or will be reported in detail elsewhere.

Tunable Laser Crystal Host Systems

H. Jenssen

Massachusetts Institute of Technology, Room 13-3146,
Cambridge, MA 02139, USA

Tunable solid state lasers were first demonstrated by Johnson, Dietz and Guggenheim as early as 1963. Only after the renewed interest in laser pumping of $Ni^{2+}:MgF_2$ and $Co^{2+}:MgF_2$ in addition to the first demonstrated tunable laser action in Cr^{3+} doped Al_2BeO_4 (alexandrite), were tunable solid state lasers considered for practical use. Since then several crystalline hosts for Cr^{3+} have been developed in addition to the demonstration of Ti^{3+} as an active ion for tunable lasers. All of these lasers belong to a class that can be loosely defined as active ion doped, optically pumped crystalline lasers. The distinction between this class and color center lasers lies mainly in the relative low stimulated emission cross section and long lifetime for the transition metal ions as compared to color centers. However d to f transitions in rare earth ions are similar to color centers.

The choice of a crystal host for an active ion tunable laser system depends on: 1. which active ion is considered; 2. mode of operation of the laser (lamp pumped or laser pumped, cw or pulsed, high average power or high peak power); 3. desired wavelength range of operation.

Ce^{3+} is an example of a rare earth ion with tunable laser potential and it has been lased in LiYF, and LaF_3. Any crystalline hosts with Y or La as one of the constituents is best suited for Ce^{3+} doping. Both charge and size of the substituting ion is a good match. It has been found, however, that this is not sufficient to guarantee lasing. In YAG excited state absorption has been identified as the reason that gain cannot be achieved. The origin of this excited state absorption, even though not fully understood, seems to be related to the coincidence of the band edge of the host and the upper 5d levels of the Ce^{3+} ion. The stonger crystal field in the oxides shifts the absorption and emission of Ce^{3+} to longer wavelengths which would be desirable for a visable tunable laser. It has also been observed that in the double perovskite fluorides ($ANaBF_6$ where A is K, Rb or Cs and B is Y, Sc or Al) the Ce^{3+} emission was shifted into the blue visible. Since fluorides generally have a wider bandgap than oxides, excited state absorption here should be of less importance. These materials are currently being studied. The experiences we have had so far from these crystals are very instructive. First, even though we have obtained crystalline samples of fair optical quality, the crystals are incongruently melting and the phase field from which growth can be done is rather limited. Second, when Nd^{3+} was used as a dopant in Rb_2NaYF_6, we have observed two different sites for the Nd ion. This is also expected to be the case for Ce^{3+}. The two sites are probably due to some interchanging of Na and Y in the lattice and therefore with a fraction of the Nd in the Na sites. These sites have nearly the same size and have the same octahedral symmetry.

The choice of a host crystal should, therefore, in addition to having a matching substituting site, not have any other sites where the active ion might fit. As mentioned above, a substitutional fit depends on the size and charge of the substituting ion, but symmetry is also important. Cr^{3+} doped garnets are interesting in this respect. $Gd_3Ga_2Ga_3O_{12}$ (GGG) has two different Ga sites, one of octahedral and one of tetrahedral coordination. When doped with Cr^{3+} the evidence is that the octahedral sites are predominantly selected. The same is true for $Y_3Al_2Al_3O_{12}$ (YAG). The coordination is, therefore, seen to be important for site selection of the Cr ion. Indeed, one of the more successful hosts for tunable laser action in Cr^{3+} is another garnet, $Gd_3Sc_2Ga_3O_{12}$ (GSGG). If more than a minute fraction of the Cr occupied tetrahedral sites, the losses would probably be too severe for laser action.

As with Ce in elpasolites, Cr doped elpasolites are plagued by poor crystal quality and also evidence of more than one site as shown by a wavelength dependent fluorescent lifetime. There is some evidence, however, that better quality crystals can be obtained where no second site has been detected. Cr^{3+} in a fluoride host is desirable for two reasons. First, the crystal field is generally weaker in a fluoride than in an oxide host. This will bring the ^4T levels responsible for the broadband emission to lower energy and thereby shift the emission to longer wavelength with less competition from the ^2E emission. Second, the product of gain and fluorescent lifetime is proportional to the inverse square of the refractive index of the host. This can typically result in a factor of two advantage for the fluorides. The disadvantage with a fluoride host is that it is typically not able to handle as high an average power as an oxide. The damage resistance is, however, at least as high as in oxides.

Ti^{3+} doped crystals for tunable lasers is the most recent entry to the field, and only sapphire has been studied in any detail. The Ti prefers to be tetravalent. A potential host crystal should, therefore, not have a site that can accomodate this ion. Al_2O_3 of course satisfies this requirement but even here it is difficult to grow the crystal with only the trivalent ion entering and post growth annealing is necessary. Fluoride host crystals might be better suited for Ti^{3+}. Fluoride crystals can be grown in a reducing atmosphere which helps retaining the trivalent titanium. As with Cr^{3+}, the weaker crystal field should also shift the absorption and emission to longer wavelengths. In sapphire the Al(Ti) is in a noninversion symmetry site. This yields a shorter fluorescent lifetime and higher cross sections than if it was an inversion symmetry site. (The same is true for trivalent Cr). This has the fortunate consequence for $Ti^{3+}Al_2O_3$ that a smaller amount of Ti is needed to achieve a sufficient absorption coefficient for efficient pumping. In garnets the site has inversion symmetry and an approximately ten fold concentration will be required. (Cr coped garnets typically have per cents Cr whereas alexandrite has tenths of per cents). The inversion site will have longer storage time however, and for lamp pumping this is important.

58

Finally, the overall efficiency of a laser material can be greatly enhanced by codoping and energy transfer. Cr sensitized Nd in GSGG and Er-Tm sensitized Ho in YLF are examples. Possible systems would include both rare earth ions and transition metal ions. Except for transfer between trivalent rare earchs and between Cr^{3+} and ND^{3+}, very little has been studied in this area. Lamp pumping of Co^{2+}, for example, might be possible if proper sensitizers can be found.

Color Center Lasers: A Brief Review

L.F. Mollenauer

AT & T Bell Laboratories, Holmdel, NJ 07733, USA

The first cw, tunable color center laser was demonstrated just eleven years ago[1]. In this talk, I shall summarize the extensive development[2] that has taken place since then in this ever-expanding field. The newly created lasers cover a large tuning range in the near infrared $(0.8 < \lambda < 4.0 \ \mu m)$, not only as cw sources capable of precise frequency definition, but also as ultrashort pulse sources as well.

The majority of laser-active color centers are based on the ordinary F center (an electron trapped at an anion vacancy), combined with some other defect in an alkali halide host. Despite their considerable variety, the various laser-active color centers have certain important features in common: all are single electron systems, tightly coupled to the lattice; hence their transitions occur as wide, homogeneously broadened bands. The lowest energy or laser transition (in both absorption and Stokes shifted emission) usually has a large oscillator strength, and gain cross-sections are high ($\sigma > 10^{-16}$ cm^2). These features are ideal for the creation of efficient, broadly tunable, high gain, loss tolerant lasers. They are also ideal for the creation of ultrashort pulses by mode-locking.

Practical laser-active color centers in the alkali halides fall into the following three categories:

F_A(II) centers: An F center perturbed by a Li ion, the F_A(II) center undergoes a qualitative change in configration following optical excitation, from the single potential well of the F center to a symmetrical well. Its pumping bands lie in the visible, while its laser tuning bands[1] are in the ~ 2.5-3.3 μm region. Stable in operation, $F_A(II)$ center lasers have already been commercialized.

F_2^+ centers: Two anion adjacent anion vacancies containing a single electron, the F_2^+ center is the solid state analog[3,4] of the H_2^+ molecular ion. The F_2^+ center is remarkable for very high efficiency and a wide range of band energies with change of host.[5-8] The pure F_2^+ center shows serious fading effects due to center reorientation with pumping, but can be stabilized by association with impurities or other stabilizing defects, as in $(F_2^+)_A$ centers[9-12] and $(F_2^+)^*$ centers.[13,14] One particular $(F_2^+)_A$ center[11] has allowed for continuous tuning over the band, so important to the spectroscopy of organic molecules, from ~ 2.5 to nearly 4.0 μm.

The $Tl^0(1)$ center: A neutral Tl atom perturbed by an adjacent anion vacancy.[15-18] The $Tl^0(1)$ center in KCl is important for the production of stable, mode-locked laser action[19] tunable in the 1.5 μm region. Lasers using $Tl^0(1)$ centers in KCl have recently become available commercially.

Cw color center lasers use a cavity configuration similar to that used for cw dye lasers. However, to avoid room temperature bleaching effects, the laser-active crystals are almost always attached to a cold finger at 77K. Thus a vaccuum chamber and a liquid nitrogen storage dewar are required. Small, inexpensive, superinsulated dewars provide nitrogen storage times of many days, and this may possibly be extended to many weeks.

Although color centers are not as suitable for high energy storage as are transition metal or rare earth ions, nevertheless, transversly pumped oscillator-amplifier combinations using $(F_2^+)_A$ centers have recently[20] produced ~10 nsec pulses of several mJ.

Because the gain medium is static, color center lasers are capable of better frequency definition than their dye laser counterparts. Recently, such definition has been

61

demonstrated[21] to better than one part in 10^{10} with a color center laser. Almost needless to say, color center lasers have proven to be excellent sources for high resolution molecular spectroscopy, especially where the molecules possess C-H and/or O-H bonds.

Mode-locked color center lasers tunable in the 1.3 and 1.5 μm regions have allowed for pioneering studies[22-25] of psec pulse propagation in optical fibers. Recently, this work has led to development of the soliton laser,[26,27] a mode locked color center laser with a length of fiber in its feedback loop. Pulse lengths, controlled by the fiber, can be as short as a few tens of fsec. This color center soliton laser is the first and only source of fsec pulses in the IR.

It should be noted that laser action tunable in the very near uv has been reported[28] with the F^+ center in the host CaO. Two very recent developments that may ultimately prove significant are the demonstration[29] of laser action (near 5 μm) with a vibrational transition (CN^- in KCl), and laser action in the visible with a color center in a gem stone[30], the H_3 center in nitrogen-doped, radation-damaged diamond.)

REFERENCES

1. L. F. Mollenauer and D. H. Olson, Appl. Phys. Lett. **24**, 386 (1974) and J. Appl. Phys. **46**, 3109 (1975)
2. A review of progress up to about 1978 can be found in: L. F. Mollenauer, "Color Center Lasers," Chapt. 6 in "Quantum Electronics, Part B," (C. L. Tang, ed.), Academic Press, N.Y., 1979
   A more complete and up to date review (same author and title) will soon be published in the "Laser Handbook", M. L. Stitch, ed. by North Holland.
   See also Sec. 2.1.3 (again, same author and title) in CRC Laser Handbook, CRC Press, Boca Raton, Fla. (1981)
3. M. A. Agerter and F. Luty, Phys. Stat. Sol. b **43**, 227f and 245f (1971)
4. L. F. Mollenauer, Phys. Rev. Lett. **43**, 1524 (1979)
5. L. F. Mollenauer, Opt. Lett. **1**, 164 (1977)
6. L. F. Mollenauer, D. Bloom, and A. M. Del Gaudio, Opt. Lett. **3**, 48 (1978)
7. L. F. Mollenauer, D. Bloom, and H. Guggenheim, Appl. Phys. Lett. **33**, 506 (1978)
8. L. F. Mollenauer and D. Bloom, Opt. Lett. **4**, 247 (1979)

9. I. Schneider and M. J. Marrone, Opt. Lett. **4**, 390 (1979)
10. I. Schneider and C. L. Marquardt, Opt. Lett. **5**, 214 (1980) and Opt. Lett. **6**, 627 (1981)
11. I. Schneider, Opt. Lett. **7**, 271 (1982)
12. I. Schneider and S. C. Moss, Opt. Lett. **8**, 7 (1983)
13. L. F. Mollenauer, Opt. Lett. **5**, 188 (1980)
14. L. F. Mollenauer, Opt. Lett. **6**, 342 (1981)
15. E. Goovaerts, J. A. Andriessen, S. V. Nistor, and D. Shoemaker, Phys. Rev. B **24**, 29 (1981)
16. W. Gellerman, F. Luty, and C. R. Pollack, Opt. Comm. **39**, 391 (1981)
17. L. F. Mollenauer, N. D. Vieira, and L. Szeto, Phys. Rev. B 27, 5332 (1983)
18. F. J. Ahlers, F. Lohse, J. M. Spaeth, and L. F. Mollenauer, Phys. Rev. B 28, 1249 (1983)
19. L. F. Mollenauer, N. D. Vieira, and L. Szeto, Opt. Lett. **7**, 414 (1982)
20. I. Schneider and C. L. Marquardt, sub. to Opt. Lett.
21. C. R. Pollack, D. A. Jennings, F. R. Petersen, J. S. Wells, R. E. Drullinger, E. C. Beaty, and K. M. Evenson, Opt. Lett. **8**, 133 (1983)
22. D. M. Bloom, L. F. Mollenauer, C. Lin, D. W. Taylor, and A. M. Del Gaudio, Opt. Lett. **4**, 297 (1979)
23. L. F. Mollenauer, R. H. Stolen, and J. P. Gordon, Phys. Rev. Lett. **45**, 1095 (1980)
24. R. H. Stolen, L. F. Mollenauer, W. J. Tomlinson, Opt. Lett. **8**, 186 (1980)
25. L. F. Mollenauer, R. H. Stolen, J. P. Gordon, and W. J. Tomlinson, Opt. Lett. **8**, 289 (1983)
26. L. F. Mollenauer and R. H. Stolen, Opt. Lett. **29**, 13 (1984)
27. L. F. Mollenauer and R. H. Stolen, Proc. of the Conf. on Ultrafast Phenomena, Monterey, Calif., June 1984 (Springer Verlag)
28. B. Henderson, Opt. Lett. **6**, 437 (1981)
29. R. W. Tkach, T. R. Gosnell, and A. J. Sievers, Opt. Lett. **10**, 122 (1984)
30. S. C. Rand and L. G. DeShazer, International Conf. on Defects in Insulating Crystals, Univ. of Utah, Salt Lake City, Utah (Aug. 1984)

An Overview of Tunable Diode Laser Technology Development

W. Lo

Physics Department, General Motors Research Laboratories, Warren, MI 48090, USA

Long wavelength ($\lambda > 22$ μm) lead-salt diode lasers are useful for spectroscopy studies and also for long distance fiber-optical communications. Double heterojunction diode lasers have now been fabricated using a new material system, $Pb_{1-x}Eu_xSe_yTe_{1-y}$. These structures were grown lattice-matched to (100) oriented PbTe substrates by molecular beam epitaxy. Laser operation up to 190K pulsed, 147K CW, has been attained with up to 1 mW single mode output power. The growth of single quantum well lead-chalcogenide diode lasers will be described. The threshold current of these quantum well lasers increases relatively slowly with temperature, yielding CW operation up to 174K (at 4.41 μm wavelength), and pulsed operation up to 260K (at 3.97 μm). To achieve single mode operation, a simple technique has been developed for the fabrication of lead-salt C^3 (cleaved-coupled-cavity) diode lasers. The improvement in spectral purity and the reduction in threshold current of these coupled cavity lasers will be discussed.

INTRODUCTION

Lead-salt diode lasers are useful for spectroscopic studies in the 2-30 μm range.[1,2] The longer wavelengths ($\lambda > 6$-8 μm) can be covered with PbSnSe or PbSnTe, and we have recently shown that PbSnTe/PbSnYbTe double heterojunction lasers can be operated CW up to 128K.[2] Shorter wavelength coverage can be obtained with PbSSe (4 μm $< \lambda <$ 8 μm) and PbCdS (2-1/2 μm $< \lambda <$ 4 μm). In the telluride system PbGeTe has been used for short wavelength coverage (3-1/2 $< \lambda <$ 6-1/2 μm) and growth of this material by molecular beam epitaxy was recently demonstrated.[3,4] A new material, PbEuSeTe, has recently been used to extend the wavelength coverage of PbTe to 4.1 μm,[5] and we now wish to report on shorter wavelength coverage with this materials system.

Figure 1 shows the development of lead-salt diode lasers in General Motors in the last 11 years. The steady increase in operating temperatures of our diode lasers over the years represents the advances in technology development in this period. With further development, room temperature operation of lead-salt diode lasers near 2 to 3 μm range seems to be possible.

HOMOJUNCTION DIFFUSED LASERS (1973-1977)

The lasers described in this section are fabricated from $Pb_{1-x}Sn_xTe$ and $PbS_{1-x}Se_x$ single crystals by diffusion technique.[6,7]

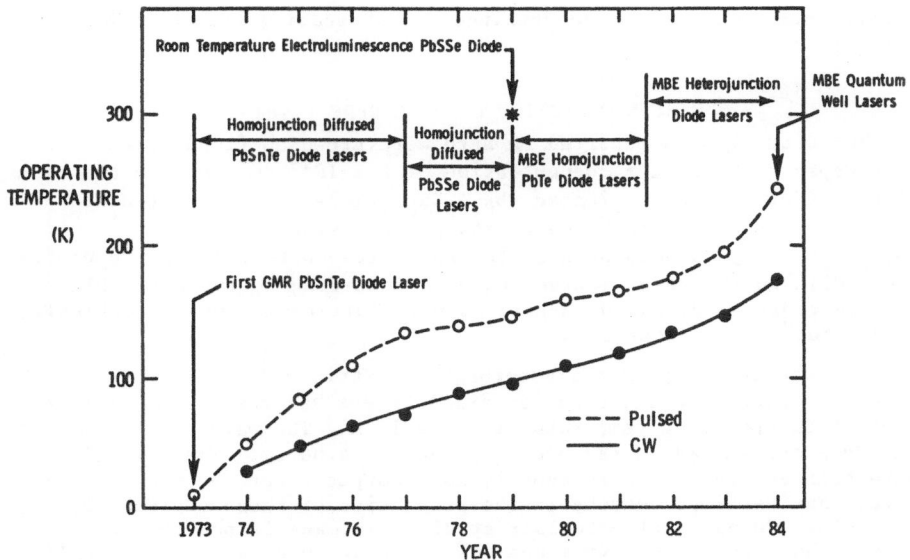

Figure 1. The improvement in operating temperatures of GM lead-salt
diode lasers in the years between 1973 to 1984

This is the most widely used technology for fabricating commercial
long wavelength (λ > 2 μm) diode lasers.

Three major areas of improvements in diode laser performance have been achieved in this period. They are: the improvement of long term reliability, the purification of mode properties,
and the achievement of a higher temperature of operation. In reliability studies we have eliminated the slow degradation caused by In
diffusion into the laser crystals.[8] By using a multilayer structure
of In-Au-Pd-Au as a contact, the room temperature storage time for our
diode lasers has been increased from months to a projection of over
10 years. In characterizing the mode properties, we found that the
lasers emit in a highly localized, filamentary manner. In order to
achieve single-mode operation, stripe widths on th order of 10 to
20 μm have been fabricated. For higher temperature operation, a
molecular beam epitaxy technique has been used to grow state-of-
the-art laser structures.[5]

MBE HOMOJUNCTION DIODE LASERS (1977-1981)

The lead-telluride diode lasers reported here operate to 115K CW,
which is 30K higher than comparable broad area diodes reported.[9] This
improvement was brought about by fabricating mesa stripe geometry
diodes with 20-35 μm stripe widths. Whereas the broad area lasers
were multi-mode except very close to threshold, these lasers exhibited
a wide range of single mode emission, especially at high operating
temperatures. Since the emission width parallel to the junction of
these lasers is about 12 μm under lasing conditions, a wider range of

single mode emission may be obtained by fabricating lasers with 10-15 μm stripe widths.

MBE HETEROJUNCTION DIODE LASERS (1981-1983)

The diode lasers reported on here were grown on (100) oriented PbTe doped with Tl at a concentration of 2×10^{19} cm^{-3} or, in the case of PbSnYbTe, on (100) oriented $Pb_{0.85}Sn_{0.15}Te$ with a carrier concentration of about 5×10^{18} due to its growth temperature.[10] $Pb_{1-x}Eu_xSe_yTe_{1-y}$ diodes have so far been grown with up to x = 0.0105, y = 0.0114 in the laser active region. According to the available lattice constant data, the MBE films were lattice-matched to the PbTe substrates within about 0.04%.

The use of europium and ytterbium in lattice-matched double heterojunction lead-chalcogenide diode lasers has resulted in attainment of CW operating temperatures up to 147K. The wavelength range of the PbTe system has so far been extended to 4.06 μm. These results have been attained with reasonably good output power and mode structure. Further improvements in the properties of these devices may be possible. Fundamental materials studies may make it possible to reduce the defect concentrations which appear to cause large tunneling currents which in turn increase the threshold current.

MBE QUANTUM WELL LASERS (1983-   )

The MBE growth system and procedures used in this study were similar to those previously reported for double heterostructure PbEuSeTe diode laser. A typical laser structure is shown in Figure 2. Other laser structure has a PbTe single quantum well active region of thickness L_z = 300 Å. The $Pb_{1-x}Eu_xSe_yTe_{1-y}$ confinement layers have x = 0.018 near the active region, yielding an increase in energy band

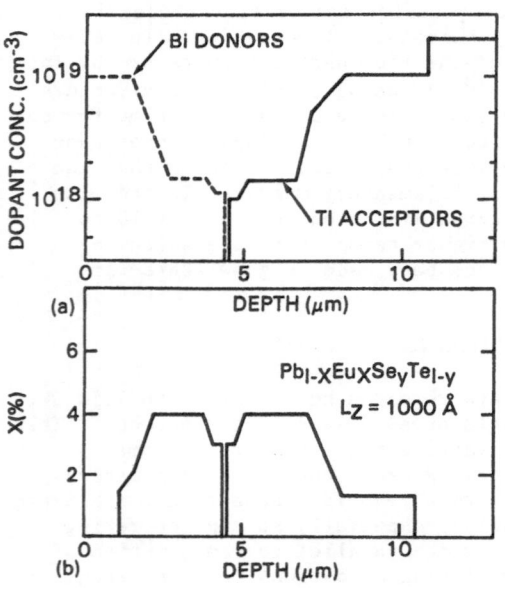

Figure 2.   (a) Dopant profile and (b) europium concentration for a laser with L_z = 1000 Å

gap of 103 meV at 80K. The europium concentration was increased farther from the active region to form a separate optical cavity structure, since the index of refraction of PbEuSeTe decreases with increasing europium concentration. Mesa stripe geometry diode lasers were fabricated as previously reported using an anodic oxide for electrical insulation. The stripe widths for these lasers were 16 to 22 μm, and the cleaved cavity lengths were 325 to 450 μm long.[11]

At low temperatures (<100 K), these lasers appear to operate on transitions between n = 1 states in the conduction and valence bands at threshold. Transitions between the n = 2 states require a higher current which decreases with increasing temperature until the laser switches to the n = 2 transition at threshold. The threshold current then increases relatively slowly with temperature. These results have enhanced the usefulness of lead-salt diode lasers for spectroscopic applications and for long wavelength (λ > 2 μm) fiber optical communications.

CLEAVED-COUPLED-CAVITY LASERS

To improve diode lasers mode property, we describe a simple technique for the fabrication of C^3 diode lasers to achieve high purity, single-mode operation from two otherwise multi-mode laser diodes.[12]

The significance of this work is two fold. First of all, we have demonstrated that a reduction in threshold current and an improvement in mode purity can be attained by simply placing a crystal reflector in close proximity to the laser cavity. If the laser is oscillating in a multi-filament mode, the feedback under constructive interference conditions is more pronounced for the strong filament and tends to result in single-filament operation. This is significant, because for lead-salt diode lasers, particularly PbSnTe lasers, single-mode operation is generally observed if single-filament operation is attained. This suggests that with the installation of a reflector in close proximity, an improvement in mode properties of lead-salt diode lasers can be obtained. Secondly, this improvement in mode purity can be further enhanced by applying an injection current to the feedback diode. Since the two diodes are cleaved from the same bar as described earlier, the filaments in the two diodes are aligned when they are pumped at the same time.

REFERENCES

1. Lo, W. and Partin, D. L., Proceedings of the Society of Photo-Optical Instrumentation Engineers, Vol. 461, pp. 5-10, 1984.

2. "Tunable Diode Laser Development and Spectroscopy Applications," Proceedings of the Society of Photo-Optical Instrumentation Engineers, Vol. 438, Wayne Lo, Editor, 1983.

3. Hesse, J. and Preier, H., Advances in Solid State Physics, H. J. Queisser, ed., Pergamon Press, 229, 1975.

4. Partin, D. L., J. Vac. Sci. Technol., Vol. 21, p. 1, 1982.

5. Partin, D. L., _Appl. Phys. Lett._, Vol. 43, p. 996, 1983.

6. Lo, W. and Swets, D. E., _Appl. Phys. Letts._, Vol. 36, p. 450, 1980.

7. Lo, W. and Swets, D. E., _Appl. Phys. Letts._, Vol. 33, p. 938, 1978.

8. Lo, W., _J. Appl. Phys._, Vol. 52, p. 900, 1981.

9. Partin, D. L. and Lo, W., _J. Appl. Phys._, Vol. 52, p. 1579, 1981.

10. Partin, D. L., Majkowski, R. F., and Thrush, C. M., _J. Appl. Phys._, Vol. 55, p. 678, 1984.

11. Partin, D. L., to be published in _Appl. Phys. Lett._, September 1984.

12. Lo, W., _Appl. Phys. Letts._, Vol. 44, p. 1118, 1984.

Part V

The Titanium Sapphire
Tunable Laser System

An Evaluation of the Ti:Sapphire Tunable Laser System

G.F. Albrecht, J.M. Eggleston, and J.J. Ewing

Mathematical Sciences Northwest, Inc., Bellevue, WA 98004, USA

NASA LIDAR applications require frequency versatile, efficient, stable, compact, long-lived laser systems for space-based applications. Systems based on Ti:Sapphire, pumped by doubled Nd:YAG lasers, are potentially well matched to this application. The attractiveness lies in its relatively broad tuning range and apparent potential for efficient operation. High system efficiencies can be projected when the Ti:Sapphire laser is excited by a frequency-doubled diode-pumped Nd:YAG laser. The diode array Nd:YAG pump technology is beginning to develop, and offers the potential for very high efficiencies at 1 μm. By using a highly tunable solid state host, instead of a dye, long lifetimes can be achieved. Stability and compactness have already been well demonstrated in well-engineered solid-state laser systems.

At MSNW we have evaluated the Ti:Sapphire laser medium for its potential as a tunable solid state laser. Two samples of Ti:Sapphire were experimentally diagnosed. The first sample (MSNW-A) was a 1/4 diameter, 1 inch long, 60° Z-cut rod with optically polished faces. The faces were Ar coated with a single layer of MgF. The rod was cored from Union Carbide[1] boule 28-943-1. The boule had 1 percent Ti:O_2 added to the melt and was 2.5 inches in diameter, 4.75 inches long at completion. The second sample (MSNW-B) was a semi-circular piece, 1/3 inch thick, 1.5 inch diameter, with an inspection polish on the semi-circular faces. The piece was orientated with the "c" axis lying in the semi-circular plane and was taken from Union Carbide boule 28-133-2. The boule had 1 percent Ti:O_2 added to the melt and was 1.5 inch in diameter, 12.7 inches long at completion. The only recorded difference between these two samples was the purity of the Al_2O_3 starting material.

The optical quality of both samples was reasonably good. There were no visible inclusions or striations in either sample and the fringe quality was about 1/4 wave at 6328 Å over a 1/4 inch diameter aperture.

Sample A was analyzed with x-ray fluorescence, and found to be uniformly doped with 0.07 weight percent Ti:O_2, or about 0.045 atomic percent titanium (2.1×10^{19} cm^{-3}). Since x-ray fluorescence is not sensitive to the valence state of the measured species, the trivalent titanium concentration is not known. No doping analysis was performed on the second sample, however its green band visible absorbance, was essentially the same as Sample B.

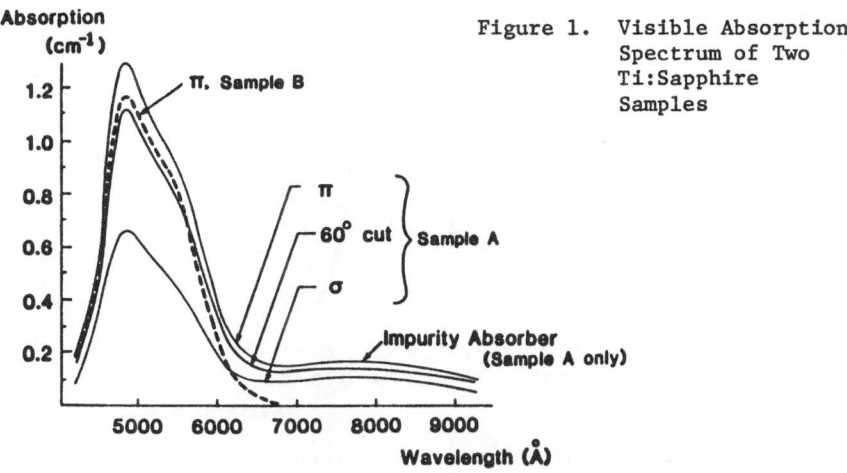

Figure 1. Visible Absorption Spectrum of Two Ti:Sapphire Samples

The measured absorption spectrum of both samples, is shown in Fig. 1. For Sample A, the π orientation absorption was deduced from the absorption measured for the σ orientation and for the 60° orientation. Sample A showed an absorption in the lasing region between 700 to 1000 nm, while measurements on Sample B revealed no absorption in this band. Our estimated detection limit for this absorption was 2 percent per centimeter. It should be noted that other investigators[2,3] have observed an absorption cross-over near 700 nm, such that the σ orientation has stronger red absorption than the π orientation. The reason for the discrepancy between their observations and ours, has not been identified. We have also observed a strong UV absorption band starting at 280 nm, which fluoresces in the blue when pumped by a KrF laser at 248 nm.

The measured lifetime of Sample A was 3.2 μsec, while that of Sample B was 3.9 μsec. If it is assumed that an impurity absorber present in Sample A causes the long wave absorbance, this impurity quenched the excited Ti^{3+} with an 18 μsec decay time. The fluorescence spectrum, calculated π cross-section and saturation fluence, under the assumption of unit quantum efficiency, are shown in Fig. 2. The observed line shape was the same for both samples, and agreed with data[4] provided by Schepler. Our detection scheme had limited sensitivity below 900 nm, thus Schepler's data was used for the red tail of the fluorescence curve shown in Fig. 2. The observed π emission was stronger than the σ emission by a factor of 1.9, which is about the same as the ratio of π to σ absorption in the green region.

From the cross-section and saturation fluence curves, the predicted system tunability range of Ti:Sapphire is 725 to 950 nm, with 700 to 1000 nm being possible if the system is designed for gain starved operation. Operation beyond these limits should be possible. However system losses and high saturation fluences will probably lead to reduced efficiencies for laser operating in the wings of the band.

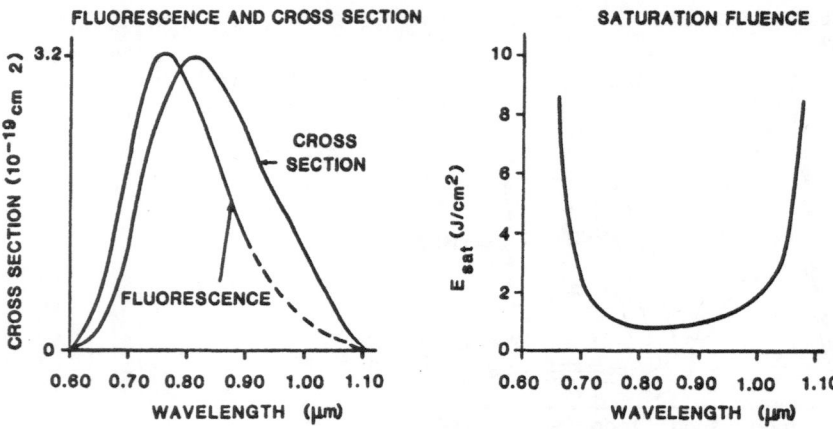

Figure 2. Ti:Sapphire Spontaneous and Stimulated
Emission Properties

By comparing lasing cross sections we see that ASE and
parasitic energy storage limitations of Ti:Sapphire should be at
least a factor of two greater than that for Nd:YAG. The peak stimulated emission cross-section of Ti:Sapphire is about a factor of
two smaller than the effective Nd:YAG cross-section, the photon energy
is slightly higher and the index of refraction is lower, which
reduces Fresnel reflections. Furthermore, it is probable that the
available sizes of Ti:Sapphire samples will be larger than currently
available Nd:YAG, which will allow even greater energy storage. Pro-
jections of short pulse red energies greater than 5 J per single
aperture are reasonable, with larger energies dependent on large
crystal sizes.

Assuming that a majority of the titanium in our samples is in
the trivalent state, the 800 nm energy storage density, at 100
percent inversion, is about 5 J/cm^3. For pump fluences greater than
4 J/cm^2, bleaching of the Ti^{3+} absorption may be easily observable.
Measuring this bleaching effect may be a way to determine the
trivalent concentration.

Laser experiments were performed with both samples. Sample
A was longitudinally pumped with 100 nsec long dye laser pulse,
and Sample B with a 8 nsec long doubled Nd:YAG laser. Optical damage
was a difficulty observed in both experiments, with a dichroic mirror
being the most susceptible to damage. The damage observed is not
thought to be intrinsic to the medium, but rather due to the presence
of hot spots in the pump beams. The observed slope efficiencies were
consistent with quantum efficiencies greater than 0.75, in agreement
with the observations of other investigators.[2] A careful measurement of this parameter is desirable, but probably not critical,
given the reasonably large value of the lower bound that has been
determined.

Transverse pump geometries, as opposed to longitudinal
geometries, used in these experiments, are desirable to reduce optical

72

damage problems. Properly polarized 532 nm pump radiation must pass
through 1.5 cm of Ti:Sapphire to be absorbed with 80 percent efficiency, at the doping densities currently available with low background loss. For a simple one-sided transverse pump geometry, with a
reflecting surface opposite the pump face, and an assumed square
extracting beam, as shown in Fig. 3, minimum amplifier face area
is 0.75 cm x 0.75 cm for good pump absorbance. For efficient extraction, the output of an amplifier must be at least 2 to 4
saturation fluences. Assuming a single extracting beam, the output
energy must be 1 to 2 J of red, which requires at least 2 to 4 J
of green to generate the needed red energy. We see that in comparison
to a transverse pumped dye laser, fairly large green pump energies
are desirable. Alternately, a low loss, higher doping density crystal,
or other pump/extraction geometries, are needed.

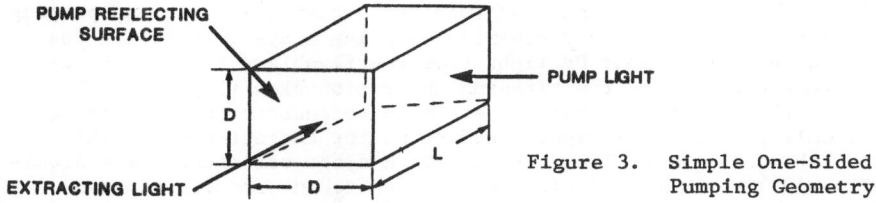

Figure 3. Simple One-Sided
Pumping Geometry

    Many properties of Ti:Sapphire still need to be determined,
before this material may become a standard material for short pulse
generation (<10 nsec). Potential properties that could greatly
impact its usefulness include: 1) low optical damage threshold,
2) upper state absorption, 3) two photon pump absorption, 4)
nonlinear index, 5) inhomogeneous broadening with a recovery time
greater than a few nanoseconds, 6) inconsistent or poor crystal
quality. It is also desirable to have large, high optical quality
samples, quickly available at a variety of doping levels.

    In conclusion, Ti:Sapphire is a promising material for an all
solid state tunable laser system when pumped with doubled Nd:YAG.
Further development is required to determine the limitations due to the
nonlinear behavior of the material, as well as careful measurements
to determine critical parameters for optimal systems design.

REFERENCES

1. A. Kokta, Union Carbide Corp., Washougal, WA.
2. P. Moulton, "Progress in Ti:Al$_2$O$_3$," NASA Workshop on Tunable
    Solid State Lasers for Remote Sensing, Stanford University,
    October 1984.
3. L. DeShazer and S. Rand, "The Ti:Al$_2$O$_3$ tunable laser system,"
    NASA Workshop on Tunable Solid State Lasers for Remote Sensing,
    Stanford University, October 1984.
4. K.L. Schepler, AFWAL/AADO, Wright-Patterson AFB, OH.

5. This work was supported by MSNW IR&D funds.

Flashlamp Pumped Ti:Al$_2$O$_3$ Laser Studies

L. Esterowitz, P. Lacovara, and R. Allen

Naval Research Laboratory, Washington, DC 20375, USA

Development of the Ti:Al$_2$O$_3$ laser[1] has mainly been confined to laser pumping, because of its short (3.2 μsec) spontaneous emission lifetime. Esterowitz, et.al.,[2] reported successful flashlamp pumping of Ti:Al$_2$O$_3$, with efficiency enhanced by using a dye surrounding the laser rod to convert near UV light from the flashlamp into blue-green fluorescent overlapping the titanium absorption band (Fig. 1). The present work describes successful attempts to substantially increase the flashlamp pumped efficiency, and to better characterize flashlamp pumped behavior. In preliminary results a slope efficiency of approximately 3x10^{-4} was obtained[2]. In our most recent results we have increased the flashpumped efficiency to 2x10^{-2}.

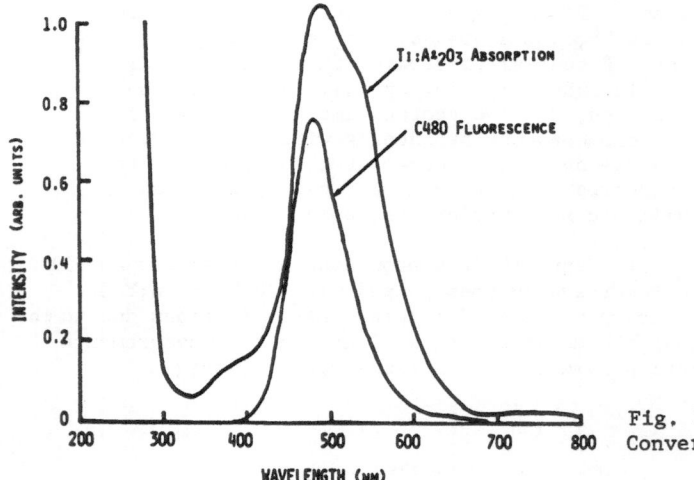

Fig. 1 Fluorescent Converter Overlap

The laser rod used in this work is 7mm x 48mm long, with AR coated ends and a pumped length of approximately 45mm. The rod was cored along the growth axis of the boule, 90° with respect to the C axis. The rod exhibits a concentration gradient along the rod axis, and slight "barberpoling". The rod is cemented into stainless steel rod holders, with the dye solution-fluorescent converter flowing in the annulus between the rod and a UV grade quartz flow tube. In addition to a fair number of scattering centers, the rod contains a small damage spot from pumping with a doubled Nd:YAG laser.

The pump cavity consists of an aluminum plated elliptical cavity, with flat aluminum plated end plates. Aluminum was chosen for its high reflectivity in the UV and blue regions of the spectrum. Because ozone created by the UV-rich flashtube emission causes rapid deterioration of the polished surfaces, the cavity is purged with dry nitrogen. Mirror spacing is 40cm, with a flat output coupler and a 5 meter concave high reflector. No attempt was made to tune beyond the tuning curve obtained in the previous work, due to a temporary lack of broader band mirrors.

A linear flashtube[3] was used with a 0.3 µF low inductance capacitor and a spark gap, with the flashtube overvoltage triggered. Excellent pulse-to-pulse stability is obtained by wrapping a grounded wire outside the envelope adjacent to the high voltage electrode. Typical flashtube inputs range from 10 to 50 Joules, with a pulsewidth of approximately 1 µsec., and a repetition rate of 0.5 Hertz.

Highest laser output efficiency has been obtained with a coumarin 480 fluorescence converter and an 80% output coupler (Fig. 2). Slope efficiencies of 0.15% are routinely recorded, with a threshold of 11 Joules. With fresh dye solution, efficiencies in excess of 0.2% have been obtained. With no dispersive elements inside the cavity, the laser output wavelength is determined by the mirror reflectivity curves and the titanium gain profile, with emission typically peaked at 788 nm, with an 8 nm FWHM. It is interesting to note that in decreasing the output coupler reflectivity from 99% to 80% the lasing threshold increased by only 2 Joules, while slope efficiency increased from 0.01% to 0.15%, indicating that the laser is undercoupled. It is likely that a further decrease in output coupler reflectivity will further increase slope efficiency.

With the flat-long radius concave resonator, and no intracavity aperture, the output mode structure appears to be a superposition of two or three low order TEM modes, with slightly distorted cylindrical

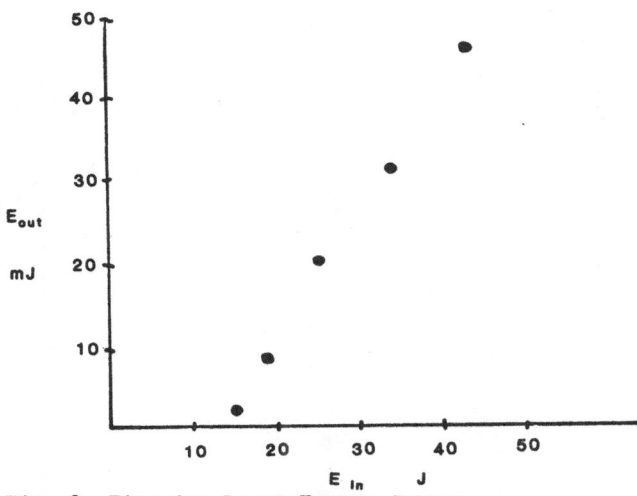

Fig. 2 Titanium Laser Energy Output

symmetry. Burn patterns in polaroid film taken at the output coupler indicate that most of the output energy is contained in a spot 2mm in diameter, with substantial energy present out to 4mm. For this resonator, the 1/e diameter of the TEM_{oo} mode is calculated to be approximately 1mm.

Our work indicates that there is considerable promise in flashlamp pumped titanium lasers. Preliminary calculations using tabulated outputs from present generation flashtubes indicate that an ultimate efficiency of about 5% could be achieved for a $Ti:Al_2O_3$ laser with an optimized fluorescent converter.

REFERENCES

1) P. F. Moulton, Laser Focus, p. 83, May 1983.

2) L. Esterowitz, R. Allen, C. P. Khattak, Stimulated Emission from Flashpumped $Ti:Al_2O_3$, presented at 1st Annual Conference on Tunable Solid State Lasers, La Jolla, CA.

3) FXQ-139C-2, EG&G, Electro-Optics, Salem, Mass.

Preliminary Experiments on Ti:Al$_2$O$_3$ Laser

L.G. DeShazer and S.C. Rand

Hughes Research Laboratories, 3011 Malibu Canyon Road,
Malibu, CA 90265, USA

We report on the laser performance of Ti:Al$_2$O$_3$ laser when pumped by a
doubled Nd:YAG laser. Several cavity configurations were used for both
60° and 90° cut Ti:Al$_2$O$_3$ laser crystals shaped as rods and plates.
Laser damage was found to be a problem with this laser when operated at
10 Hz. The scattering losses and absorption at 750 nm are being investigated, and may be due to divalent titanium ions.

For 95% output coupling in a cavity containing a 5mm x 60mm laser rod
and pumped with the extraordinary polarization at 532 nm, we obtained
1% efficiency laser operation at 750 nm. The geometry used was far
from optimum. No attempt to match modes was made and damage occurred
when pump beam power was raised only slightly above threshold at 10 Hz.
Damage occurred to mirrors in the longitudinal pump geometries and to
the Ti:Al$_2$O$_3$ sample especially when pumped with the ordinary polarization. Progressive asterization was also noticed prior to damage in the
ordinary polarization.

Ti:Al$_2$O$_3$ Laser Pumped
by a Frequency Doubled Nd:YAG Laser

N.P. Barnes and D.K. Remelius

Los Alamos National Laboratory, P.O. Box 1663,
Los Alamos, NM 87545, USA

A Ti:Al$_2$O$_3$ laser was developed that was capable of producing 5.4 mJ of output with a slope efficiency of 0.3. The primary limitation to this laser was caused by the relatively large loss in the Ti:Al$_2$O$_3$ crystal at the lasing wavelength, about 0.16 cm^{-1}. This high loss both caused the high threshold and limited the slope efficiency. The laser was characterized by measuring the laser output and the pulse length as a function of the energy input. The former measurement was also done as a function of the output mirror reflectivity. From this data, the slope efficiency and the threshold of the laser was determined as a function of the mirror reflectivity.

The Ti:Al$_2$O$_3$ laser was modeled and the results of the model were compared with the experiment. The modeling could be relatively accurate since not only could the total stored energy be readily determined but also the radial profile of the stored energy could be determined. The results of the modeling agreed quite well with the experimental data.

The experimental arrangement is shown in Fig. 1. The Nd:YAG oscillator could produce 50 mJ of energy in a beam which was a little

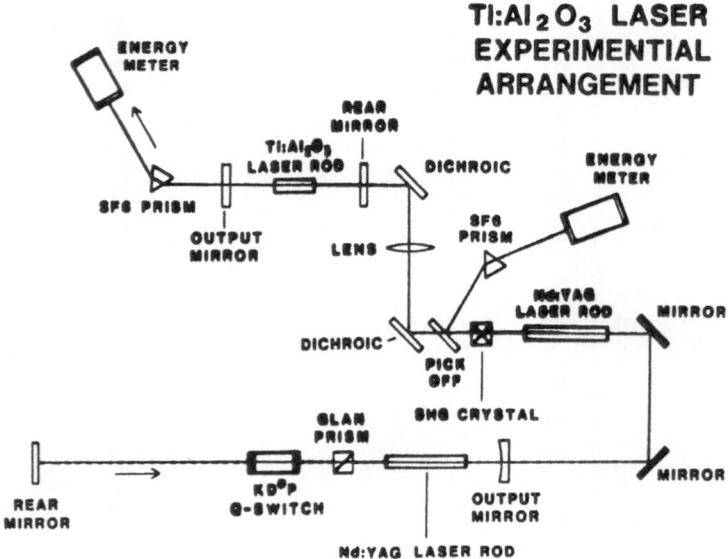

Fig. 1

under twice diffraction limited. The pulse length of the laser was approximately 15 ns. The beam was then amplified to 150 mJ by a single 6.3 by 76-mm amplifier. The output of the amplifier was frequency doubled to produce 25 mJ of second harmonic in a 30-mm KD*P crystal using Type II phase matching. Later, a 30-mm CD*A crystal was used that could produce 50 mJ of second harmonic, however, it did not arrive in time to be used for the experiments. A glass plate was used to sample the second harmonic energy. After separating the fundamental from the second harmonic with a SF_6 prism, the sample was measured on an energy monitor. The fundamental in the main beam was separated by two dichroic mirrors which were highly reflecting at 0.532 μm. The second harmonic was focused into the $Ti:Al_2O_3$ with a long focal length lens through a dichroic mirror. The $Ti:Al_2O_3$ resonator consisted of a dichroic mirror having a 5.0-m RC, an 18-mm $Ti:Al_2O_3$ sample with AR coated surfaces, and a flat output mirror. The unabsorbed 0.532-μm beam was separated from the $Ti:Al_2O_3$ output with a second SF_6 prism. The laser output energy was measured using 100 shot averages.

The energy profile of the second harmonic was found to be approximately Gaussian distributed. The profile was determined by measuring the energy transmitted through an aperture as a function of the aperture diameter. A plot of the data showed that the distribution appeared to be well approximated by a Gaussian and a curve fitting procedure yielded the beam radius at the position of the $Ti:Al_2O_3$ crystal.

The $Ti:Al_2O_3$ crystal was characterized to provide data for the model. The upper laser level lifetime was found to be 3.2 μs. The absorption of the second harmonic was measured at 0.74. The reflection loss of the 0.532 μm at the surface was less than 0.01 while the reflection loss at 0.788 μm was less than 0.002. However, the bulk loss at 0.788 was measured to be 0.16 cm^{-1}.

As the energy profile of the second harmonic was known, the gain could be accurately determined. The small signal gain of a TEM_{oo} mode, G_A, was approximated as

$$G_A = \int_0^\infty (2/\pi w_0^2) \exp(-2\rho^2/w_0^2) \exp[\sigma_e \ell N_R \exp(-2\rho/w_p^2)] 2\pi\rho \, d\rho$$

In the expression, w_0 is the pump beam radius σ_e is the effective stimulated emission cross section, ℓ is the length of the crystal, N_R is the peak population density in the upper laser level, and w_p is the pump beam radius. This expression can be evaluated exactly in two cases, namely where $w_0 \longrightarrow 0$ and where $w_0 = w_p$. In other instances, numerical techniques can be used. The results of the numerical calculations appear in Fig. 2. With a knowledge of the small signal gain, the threshold can be readily computed. This, in turn, allows the laser output energy to be calculated as a function of the pump energy using standard laser theory and knowing only a few other readily measurable parameters, such as the output mirror reflectivity. Knowing the stored energy also allows the prediction of the laser output energy as a function of the laser output energy.

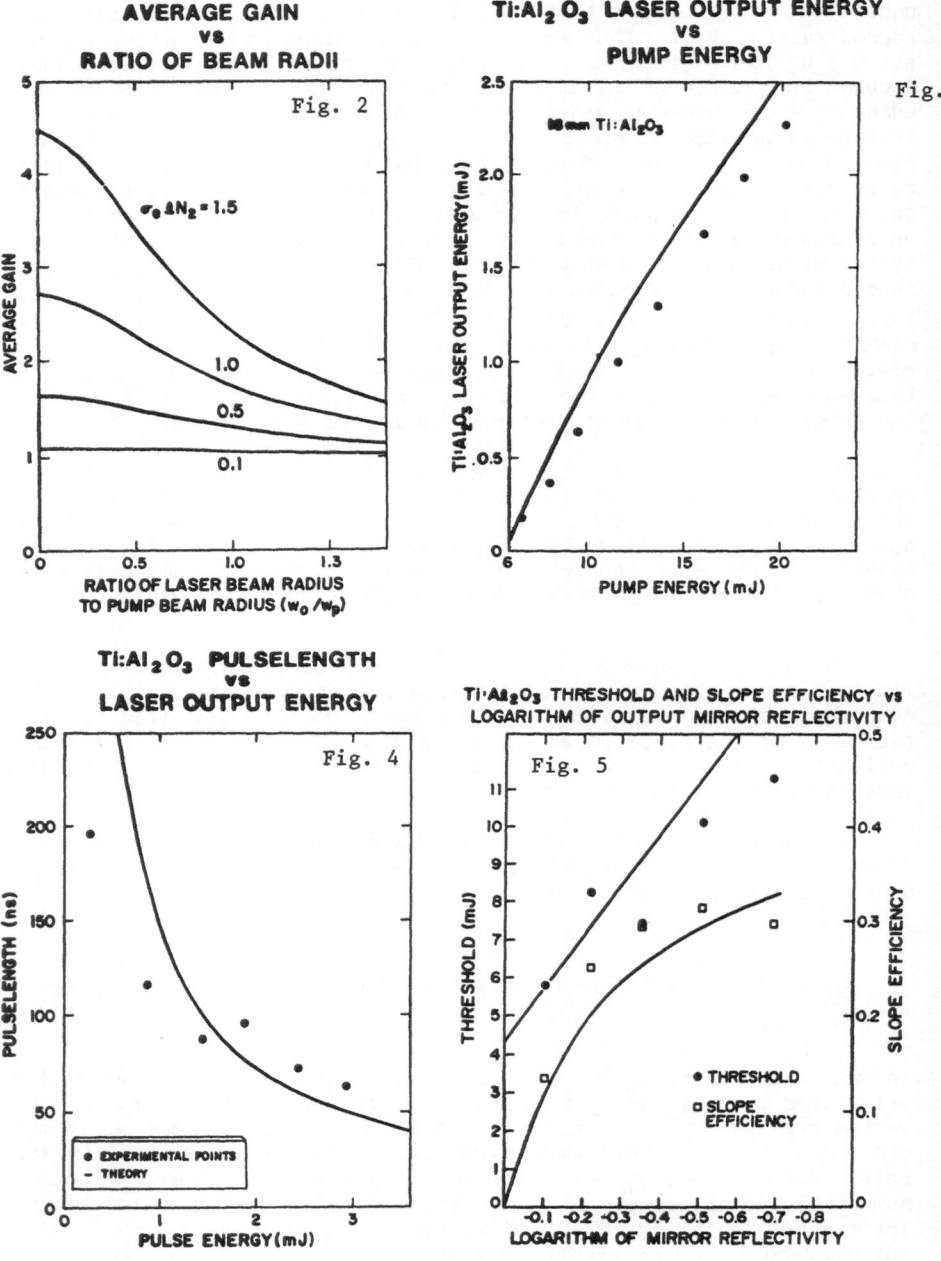

AVERAGE GAIN vs RATIO OF BEAM RADII

Fig. 2

AVERAGE GAIN

$\sigma_0 \Delta N_2 = 1.5$

1.0

0.5

0.1

RATIO OF LASER BEAM RADIUS TO PUMP BEAM RADIUS (w_0/w_p)

Ti:Al$_2$O$_3$ LASER OUTPUT ENERGY vs PUMP ENERGY

Fig. 3

88 mm Ti:Al$_2$O$_3$

Ti:Al$_2$O$_3$ LASER OUTPUT ENERGY (mJ)

PUMP ENERGY (mJ)

Ti:Al$_2$O$_3$ PULSELENGTH vs LASER OUTPUT ENERGY

Fig. 4

PULSELENGTH (ns)

● EXPERIMENTAL POINTS
— THEORY

PULSE ENERGY (mJ)

Ti:Al$_2$O$_3$ THRESHOLD AND SLOPE EFFICIENCY vs LOGARITHM OF OUTPUT MIRROR REFLECTIVITY

Fig. 5

THRESHOLD (mJ)

SLOPE EFFICIENCY

● THRESHOLD
□ SLOPE EFFICIENCY

LOGARITHM OF MIRROR REFLECTIVITY

The experimental results show good agreement with the predictions of the model. These results appear in Fig. 3 for one mirror reflectivity. Good agreement can be seen, for example at the highest input energy, the measured slope efficiency is 0.30 while the predicted

80

slope efficiency is 0.31. In Fig. 4, the results of the laser pulse-length as a function of the output energy measurements appear with the predictions of the model. In this case the agreement is not quite as good, however the general trend is still predicted well. The stimulated emission cross section measured by Moulton, 2×10^{-19} cm^2, was used in these calculations. Since only one figure accuracy is quoted, part of the disagreement may be rectified with a more accurate value for this parameter.

The laser performance was also characterized as a function of the output mirror reflectivity. The laser output energy was measured as a function of the pump energy for each output mirror used. This data was then reduced to yield a threshold and a slope efficiency. The slope efficiency and threshold were then plotted as a function of the logarithm of the output mirror reflectivity. The results appear in Fig. 5. The threshold and the slope efficiency, as predicted by the model, are also shown in the figure. Again, reasonable agreement can be found between the experimental measurements and the predictions of the model.

The present Ti:Al$_2$O$_3$ laser was limited by the high loss associated with the particular laser crystal and by the limited amount of pump energy. Further crystal development should eliminate the high loss problem. The limited pump energy problem has already been mitigated by merely employing a CD*A doubler. Consequently, even better performance can be expected from Ti:Al$_2$O$_3$ in the future. This material should find application much as dye lasers pumped by frequency doubled Nd:YAG lasers do. In addition, Ti:Al$_2$O$_3$ begins to perform well in a spectral region where the performance of dyes are beginning to decrease.

Absorption and Fluorescence of Alexandrite and of Titanium in Sapphire and Glass

C.E. Byvik[1], *A.M. Buoncristiani*[2], *and R.V. Hess*[1]

[1]NASA Langley Research Center, Hampton, VA 23665, USA
[2]Christopher Newport College, Newport News, VA 23606, USA

INTRODUCTION

An assessment of ions and hosts as candidates for new laser materials can be made from absorption and fluorescence measurements. Predictions of important laser parameters such as gain, absorption, and emission cross-sections can be readily made from this data. Titanium in crystalline sapphire exhibits a broad fluorescence spectrum with a peak near 750nm. A search for new laser host materials for Ti(III) has shown that Ti:YAlO$_3$ also exhibits a broad fluorescence spectra with the peak blue shifted by about 150nm (ref. 1). We undertook a study of Ti(III) in glass hosts which should exhibit an extreme low field case for the ion (ref. 2). We present here the results of fluorescence and absorption experiments made, to date, on titanium in crystalline sapphire, at two dopant levels, and titanium doped into three glass structures and compare these results with those of alexandrite.

RESULTS

The absorption and fluorescence experiments were carried out using samples consisting of crystalline sapphire and glasses doped with Titanium (III). No attempt was made to observe the effects of the polarization of radiation. The doping levels and host materials are summarized in Table 1. The doping concentrations are the values stated by the suppliers. For the glasses, the doping concentration is the weight percent of Ti$_2$O$_3$ added to the melt.

The Titanium doped silicate glass samples were clear to the eye and each exhibited no absorption attributable to the Ti(III) ion. (Attempts were made by the supplier to prevent the oxidation of the Ti(III) in the melting process.) The Titanium doped phosphate glass, however, is deep blue in appearance and its absorption spectrum is shown in Figure 1. Also shown in Figure 1 is the absorption spectrum for the 0.09% Titanium in sapphire. The spectrum for the Titanium in glass has a peak at 570nm which is shifted approximately 100nm to the red from the Ti:Sapphire absorption peak. The knee on the low energy side of the peak attributed to the Jahn-Teller effect is observed in both spectra. It should be noted that the shift in the peak of the absorption spectrum illustrates the sensitivity of the Ti(III) ion to the crystalline environment or lack of it in the case of the glass host. Furthermore, the glass sample exhibits an absorption linewidth of about a factor of 2 larger than that of sapphire. This is attributed to inhomogeneous broadening expected for the Ti(III) in the

Table 1.

Table 1.

Titanium Host Material	Ti$_2$O$_3$ Doping Concentration	Absorption Peak(nm)/Linewidth(cm^{-1})	Emission Peak(nm)/Linewidth(cm^{-1})
Sapphire	0.09%	475/5150	750/2900
Sapphire	0.2 %	Not Measured	790/3000
Silicate Glass (14.6% Al$_2$O$_3$)	1.0 %	None	None
Silicate Glass (11.7% Al$_2$O$_3$)	1.0 %	None	None
Phosphate Glass (7.0% Al$_2$O$_3$)	1.0 %	570/9650	Undetermined

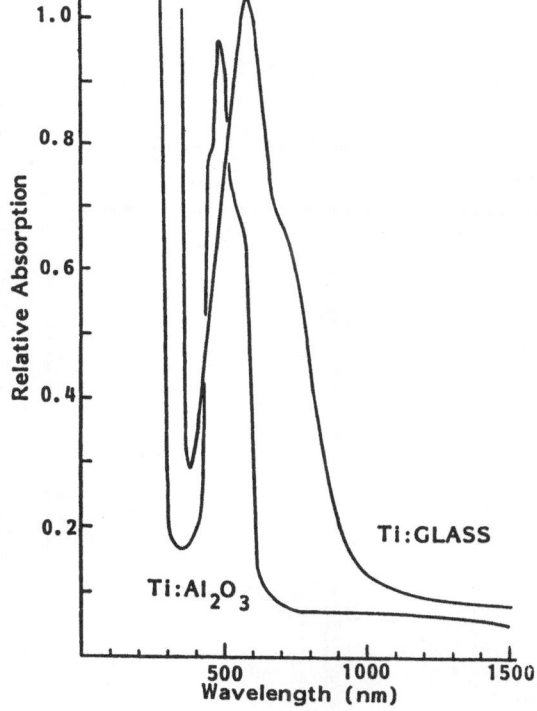

Figure 1. The relative absorption spectra of Titanium doped glass and sapphire

glass host, a broadening which would degrade the performance of this material as a laser medium.

Fluorescence measurements were made using the 488nm line of the Argon-ion laser. Figure 2 shows typical photoluminescence spectra for Titanium doped sapphire and alexandrite. The fluorescence spectra exhibits significant structure compared to that of the Ti:Al$_2$O$_3$. The two large and narrow peaks on the high energy side are the zero phonon transitions (R lines) of the chromium ion. The zero phonon transitions

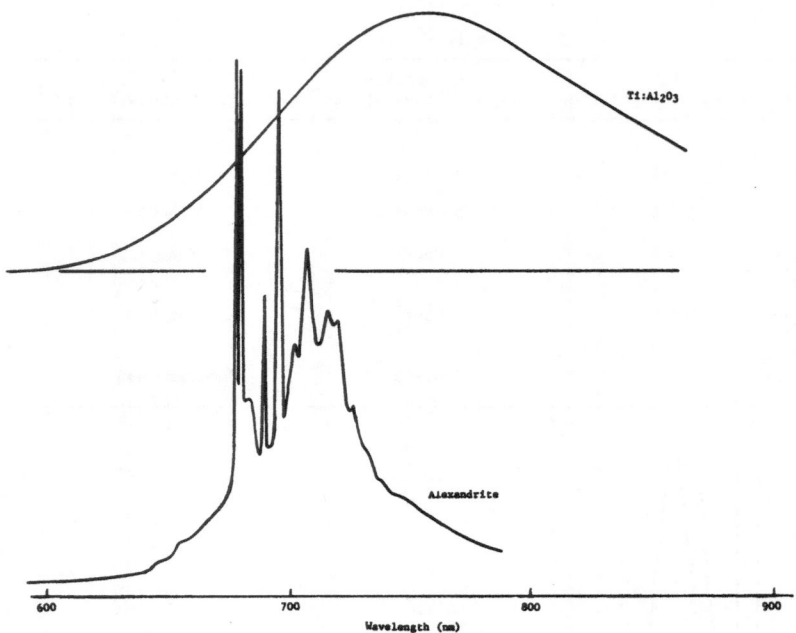

Figure 2.   Comparison of the photoluminescence spectra of Ti:Al$_2$O$_3$
            and Alexandrite.

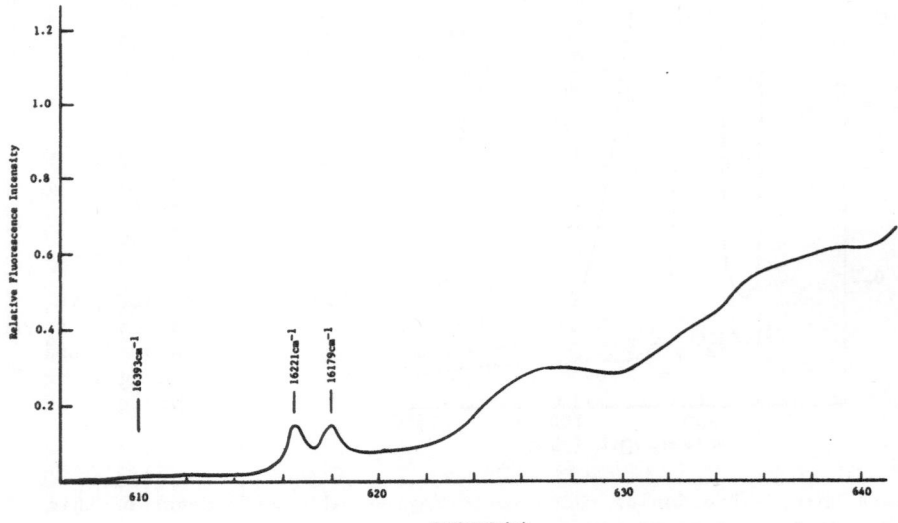

Figure 3.   Fluorescence spectrum of Ti:Al$_2$O$_3$ at liquid nitrogen
            temperatures showing two of the three zero phonon
            transitions as well as single phonon transitions.

for the $Ti:Al_2O_3$ are not observed in the fluorescence spectra at room temperature but two of the three transitions can be observed at liquid nitrogen temperatures (see Figure 3). It is apparent from Figure 2 that the Ti:Sapphire will lase over a potentially broader range than the alexandrite. The fluorescence spectrum for the 0.2% $Ti:Al_2O_3$ is similar to that shown in Figure 2, however, structure is observed near the peak. Also, the R-lines of chromium are observed in all of the Ti:Shapphire fluorescence spectra.

As mentioned previously, two of the three zero phonon transitions for $Ti:Al_2O_3$ can be observed at liquid nitrogen temperatures. This structure in the fluorescence spectra is shown in Figure 3. The two peaks at $16221cm^{-1}$ and $16179cm^{-1}$ correspond to transitions from the ^2Eg level to the $E_{3/2}$ and the $_1E_{1/2}$ states, respectively. The third zero phonon transition (from the ^2Eg to the $_2E_{1/2}$ state) is not observable at this temperature and is expected to be observed at liquid helium temperatures on the low energy side of the two sharp peaks. The phonon assisted transition at $16393cm^{-1}$ is attributed to transitions from the first and second phonon states of the ^2Eg electronic state to the zero phonon state of $^2T_{2g}$. The estimated energy for the inter-acting phonon with the ^2Eg electronic state is $172cm^{-1}$. The structure on the low energy side of these zero phonon peaks are attributed to transitions to phonon states in the $^2T_{2g}$ ground state.

The 488nm and 514.5nm lines of the Argon ion laser as well as the 632.8nm line from a helium-neon laser were used as sources in fluorescence experiments with the Titanium doped phosphate glass. No fluorescence was observed out to the wavelength limit (880nm) of the detector. Experiments are underway to determine if there is fluorescence beyond 880nm. A search for other host materials for Ti(III) to provide a widely tunable solid state laser medium at both longer and shorter wavelengths than Ti:Sapphire is continuing.

[1] P. F. Moulton, Private Communication.
[2] P. T. Kenyon, L. Andrews, B. McCollum and A. Lempici, IEEE J. Quant. Elect., Vol. QE-18, 1189 (1982).

Part VI

Advanced Laser Concepts –
The Slab Geometry

Progress in Solid-State Slab-Lasers

Y.S. Liu and W.B. Jones

General Electric Company Research and Development Center,
Schenectady, NY 12301, USA

Progress of solid-state slab-geometry lasers and their applications to the development of tunable laser sources are reviewed.

The advance of slab-geometry solid-state lasers has led to much of the recent progress in the development of high-average-power, frequency-agile laser sources. (1) The solid state laser with a slab geometry has been designed for generating a near-diffraction-limited beam at high average power. They are ideal pump sources for many nonlinear frequency conversion applications using harmonic generation and/or stimulated Raman scattering processes. Recently, there have been several developments of novel broadband tunable solid-state-laser materials and the efficient second-harmonic nonlinear crystal KTiOPO4 (2). With these concurrent developments, the slab-type configuration provides a particularly attractive approach for developing a solid-state-laser-based tunable coherent source which can be continuously tunable across the entire UV to IR spectral range and operated at a high-average-power level. We will review the progress in solid-state slab-lasers and their applications to the development of tunable laser sources using various nonlinear optical techniques.

Conventionally, the solid state laser has been fabricated in the form of a rod shape optically pumped with a flashlamp. Under a thermal loading condition, the thermally induced focusing and birefringence caused by the nonuniform temperature distribution inside the gain medium limits the beam quality as well as the output power. In the slab-type laser configuration, the solid host material (e.g. Nd:glass, Nd:YAG) is fabricated in the form of a rectangular cross-section slab with plane parallel surfaces, shown schematically in Fig. 1. Under the ideal condition in which the slab is simultaneously pumped and cooled through two opposite slab faces, a one-dimensional, steady-state thermal distribution is achieved, and the temperature gradient is normal to the slab faces. Under this condition, an optical beam propagates through a zig-zag path via total-internal-reflection between the two slab faces sees no net thermal lensing effect; a

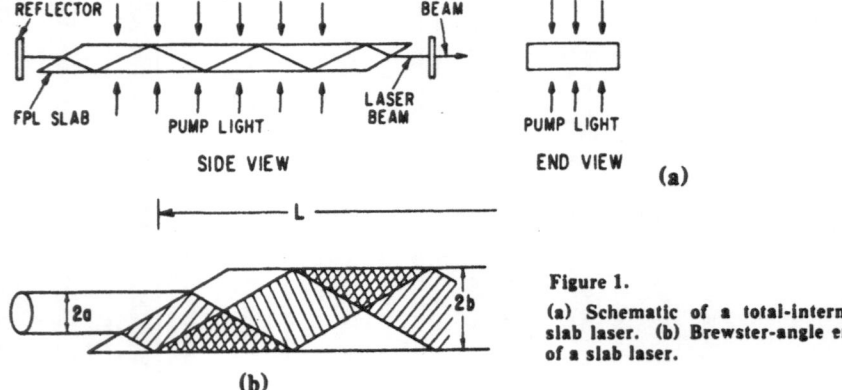

Figure 1.

(a) Schematic of a total-internal-reflective slab laser. (b) Brewster-angle entrance face of a slab laser.

result that is easily concluded from a symmetry point of view. In addition, thermal-induced stress is averaged zero for a beam passing from one face to another face such that no depolarization occurs.

A slab laser device, however, may deviate from the ideal situation due to nonuniform pumping and cooling, edge and end effects, and effects due to various boundary conditions. For example, when the inversion pumping is not uniform across the slab width, transverse thermal gradients will occur for which there is no intrinsic compensation.Consequently, care must be taken that the pumping distribution is uniform across the slab width to avoid distortion in the direction parallel to the width. In practice, this entails a careful design of the reflectors for flashlamp pumping, as well as efficient coupling between the flashlamp and the gain material. Equally important, slab cooling must be designed such that the faces are cooled to equal temperatures and temperature gradients along the slab faces are minimized. In addition, a slab laser device is further limited by those constraints such as thermal fracture limits, optical damages and spontaneous amplifier emission among others.

The slab configurations have been operated in a variety of modes for a number of applications. In all designs that have been implemented, the pump lamps have been linear discharge flashlamps Kr-filled for CW operations and Xe-filled for pulsed operations. Both Nd:YAG and Nd:glass materials have been fabricated into slab forms and operated as an oscillator and/or amplifier. Single slab Nd:YAG laser designs have operated CW at 150 Watts average output power, normal pulse mode at 5 J/pulse and 20 pps, and Q-switched at 0.5 J/pulse, 20 pps, all with beam quality less than five times diffraction-limited. Designs with Nd:YAG have also been developed as low-distortion amplifiers for Q-switched oscillators at high repetition rates. A single Nd:YAG slab device, designed and operated as Q-switched oscillator and multi-pass folded-amplifier, has produced over 80 Watts average power at a repetition rate over 300 Hz. A doubling efficiency of 40% from the output beam of this slab device has been achieved. This example demonstrated a slab laser can be operated at a high power level while maintaining an excellent beam quality.

In contrast to Nd:YAG , Nd:glass is available in essentially any dimension with excellent optical quality. The specific gain coefficient of this medium makes it useful only for pulse operation. The low heat conductivity is the key factor that sets a limit on glass slab thickness that can be used in a given design without exceeding the thermal fracture limit. In typical designs for operation in the range to 10 pps, the inversion that can be obtained, which is compatible with useful flashlamp pumping conditions, falls in the range of 0.25 to 0.5 J/cm3 per pump pulse. The principal application of the Nd:glass slab laser has been as a Q-switched oscillator-amplifier in a single slab operated at a repetition rate to 10 pps with an output energy to over 2.5 J/pulse. The beam quality is sufficiently good for efficient second harmonic generation in deuterated KDP crystal.

Another application of slab-type glass laser is the generation of high-peak-power short pulses of nanosecond and picosecond durations at a high-average-power output. We have applied the "intracavity injection" technique to a Nd:glass slab oscillator to generate optical pulses with durations of one nanosecond to several picoseconds at a repetition rate to 10 pulse- per-second. (3) Extremely stable and well synchronized short pulses were achieved in these devices with an output pulse energy limited only by the damage of optical components such as Pockels cell and dielectric coating inside the oscillator. In conjunction with a slab configuration, the intracavity injection technique should be applicable to a variety of wideband solid state laser materials to generate both spectrally and temporally tunable output over a wide range of pulse widths at a high-average-power level, a unique property which would be difficult to achieve by any other means.

Unlike the conventional rod geometry, the slab laser with its large aspect ratio lends itself to incorporation of both oscillator and amplifier functions in a single laser head, thus providing a higher overall efficiency with a significant reduction in system complexity. Although the thermal self-compensating aspects of the slab geometry and the low gain necessary for the specific performance would permit the operation of an oscillator at high average powers, other optical components such as Pockels cell are more limited in their thermal capability for handling high average power. Phosphate glass such as LHG-5 and Q-88 were used for their good optical as well as mechanical properties. Some of the slab laser developments and recent status will be reviewed.

REFERENCES:

1. Y.S. LIU, W.B. JONES and J.P. Chernoch, *Proc. of Special Topics in Optical Propagation,* P. Halle, ed., (Advisory Group for Aerospace Research and Development, NATO, Paris, 1981) pp. 301-309; also G.E. Technical Information Series, Rep. No.81CRD104, (General Electric, Schenectady, New York, 1981)

2. Y.S. Liu, L. Drafall, D. Denz, and R. Belt, "Nonlinear Optical Properties of KTiOPO4," in *Technical Digest of Conference on Lasers and Electro-Optics,* (Optical Society of America, Washington, D.C. 1981), p.26.

3. Y.S. Liu and W.B. Jones, "Temporally Tunable Q-switched Nd:Glass Slab Laser Using Intracavity Self-Injection and Nonlinear Pulse Compression Techniques" in *Technical Digest of Conference on Lasers and Electro-Optics,* (Optical Society of America, Washington, D.C. 1983), p.154.

Scalability of High Average Power Solid State Lasers

D.C. Brown

General Electric Company Corporate Research and Development, Schenectady, NY 12345, USA

ABSTRACT

There is much current interest in the realization of high average power (HAP) solid state lasers at the kilowatt level and beyond. In this paper we review laser devices and system architectures that may be used to reach such power levels, their scaling laws, and review the physics and technology issues that must be taken into account. A comparison of the power handling capability of Nd:LHG-5, Nd:YAG, CR:Alexandrite, and Ti:Sapphire is also shown.

Current HAP solid state laser devices typically operate with power outputs of tens of watts with near diffraction-limited performance. Slab laser devices with outputs of hundreds of watts have also been operated with greater beam divergence, while rod geometry lasers have been operated at the kilowatt level with poor beam quality. A wide variety of commercial and military laser applications require laser devices with HAP output of at least a kilowatt, with close to diffraction-limited beam quality. While slab laser devices clearly have the capability of operating in that regime, it is important to begin to understand the limitations of that approach and to consider what devices might be suitable as scalable HAP amplifiers. In Figures 1 and 2, we show schematics of a number of amplifier geometries. In the slab laser geometry, three different configurations are shown. The straight-through case is currently in use in the Soviet Union DELPHYN laser system, as well as in a commercial laser system. The Zig-Zag case, invented and under continual development by the General Electric Company, is the most common arrangement, since in the ideal case, thermally induced birefringence vanishes and thermal focusing is compensated to first-order. In the sandwich configuration, which has a substantially higher efficiency than the ordinary slab laser (1), two slabs are pumped in a back-to-back fashion; thermally induced bowing is eliminated by adjusting the cooling conditions on the two cooled faces to give a uniform thermal profile. The slab laser approach is favored when compared to rod lasers due to the higher obtainable beam brightness. There are a number of disk configurations of interest (see Fig. 2), including the salami or zero axial gradient laser consisting of a multiplicity of laser disks. This arrangement has not found wide favor in the past because of the lack of good index matching fluids leading to unacceptably large insertion losses. Radial energy deposition also lead to incompensatable thermally-induced focusing. In the ordinary disk laser where the disks are mounted at Brewster's angle, two additional transparent plates allow the disks to be actively cooled by a fluid. This configuration is under consideration for HAP systems at Lawrence Livermore Laboratory using flowing He or N_2 as the coolant, and is capable of high power operation. In order to achieve uniform illumination of the disks and thus minimize thermal distortions, it is necessary to consider other geometries. For example, the prism disk laser shown in Fig. 2 results in uniform illumination (2). Similarly, the uniform illumination configuration (3) in Fig. 2 achieves the same purpose. The prism disk laser has been employed in the past in the Alpha system constructed at the Laser Fusion Feasibility Project at the University of Rochester, but was discarded due to its large pathlength in glass, leading to the onset of small-scale self focusing for picosecond pulses, due to the finite nonlinear index of refraction. For Q switched and normal mode operation, however, this device deserves consideration as a HAP amplifier. Another promising amplifier, shown in Fig. 3, is the active-mirror which may also be operated in the high efficiency sandwich configuration. Recently, such a device was con-

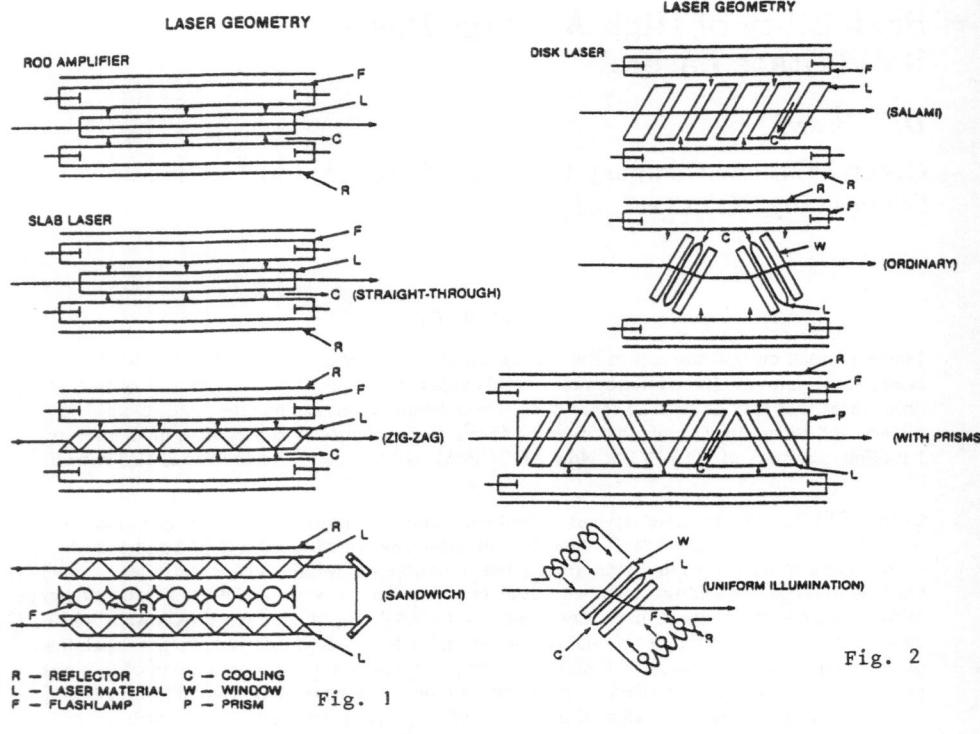

LASER GEOMETRY

ROD AMPLIFIER

SLAB LASER

(STRAIGHT-THROUGH)

(ZIG-ZAG)

(SANDWICH)

R — REFLECTOR   C — COOLING
L — LASER MATERIAL   W — WINDOW
F — FLASHLAMP   P — PRISM

Fig. 1

LASER GEOMETRY

DISK LASER

(SALAMI)

(ORDINARY)

(WITH PRISMS)

(UNIFORM ILLUMINATION)

Fig. 2

LASER GEOMETRY

ACTIVE-MIRROR

(ORDINARY)

(SANDWICH)

Fig. 3

structed and operated; over 120W of extractable power was demonstrated at a repetition rate of 5 Hz (4). Thermally induced focusing may be completely eliminated, as in the sandwich slab laser, by actively cooling both faces and adjusting the cooling conditions to be slightly different.

In designing HAP systems, a number of system architectures may be considered, as shown in Fig. 4. The single-pass system is most often used today, but not capable of as large a system extraction efficiency at the double, multiple pass, on regenerative systems. The multiple pass system is commonly used to extract energy from slab lasers (5). The achievement of HAP regenerative systems has been and continues to be hampered by the lack of a scalable optical switch. Another promising approach to scaling up involves the use of the phased-array concept where small, highly efficient lasers are locked together by some appropriate process. In

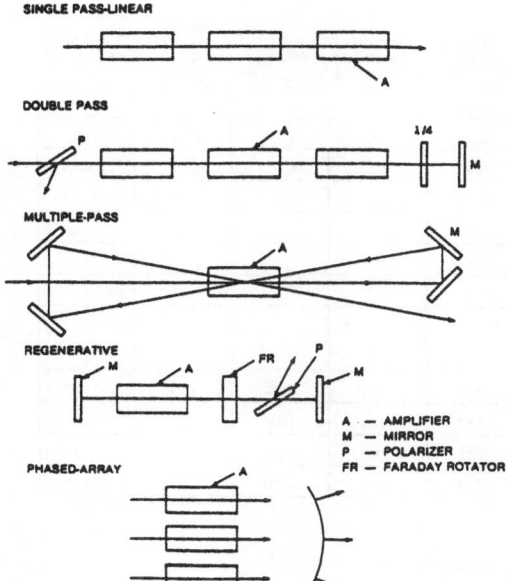

SYSTEM ARCHITECTURES   Fig. 4

SINGLE PASS-LINEAR

DOUBLE PASS

MULTIPLE-PASS

REGENERATIVE

A — AMPLIFIER
M — MIRROR
P — POLARIZER
FR — FARADAY ROTATOR

PHASED-ARRAY

SCALING LAW LIMITATIONS

- OPTICALLY INDUCED DAMAGE

- PUMP LIMITATIONS

- AMPLIFIED SPONTANEOUS EMISSION

- PARASITIC OSCILLATIONS

- NONLINEAR INDEX OF REFRACTION

- MATERIAL LIMITATIONS

Fig. 5

the far field, the resulting wavefront appears to be coherent emission from a single large aperture.

Simple, approximate scaling laws have been worked out for common laser geometries and are summarized in Fig. 5. For the sandwich configurations (disk, slab, active-mirror), the ordinary maximum average power at the material fracture limit P_{AV}^M should be multiplied by 2, owing to the increase in laser volume. These laws may be summarized by the equation

$$P_{AV}^M = K \left[\frac{\eta_{ex}}{\chi} \right] R_m \left[\frac{V}{t^2} \right] = K \left[\frac{\eta_{ex}}{\chi} \right] R_M \left[\frac{A_p}{t} \right]. \tag{1}$$

Here η_{ex} is the extraction efficiency, χ the ratio of peak inversion density to pulse integrated heat density, V and A_p the laser glass volume and pumped area respectively, and t the thickness. K is a geometrical constant whose values are given in Table 1.

Table 1

Amplifier	K
Rod	$8\pi / \sqrt{2}$
Slab	$12 / \sqrt{2}$
Disk	$3\pi / \sqrt{2} \, (n_o^2 + 1)^{1/2}$
Active-Mirror	$6\pi / \sqrt{2}$

Note: For sandwich configurations multiply by 2.

The constant R_M known as the figure of merit, rupture modulus or thermal resistance parameter, is given by:

93

Fig. 6

Crystals	Thermal Conductivity k (w/cm-°C)	Poisson's Ratio ν	Thermal Expansion Coefficient α (1/°C)	Young's (E) Modulus (kgm/cm²)	Material Constant m_s(w-cm/kgm)	Strength σ_m (kgm/cm²)	Material Figure of Merit R_s(w/cm)	Normalized Figure of Merit
	(300 °K)		× 10⁻⁶	× 10⁵	× 10⁻⁴	× 10³		
Y₃Al₅O₁₂ (YAG)	0.13	0.252	7.04	27.30	50.6	2	10.12	29.61
BeAl₂O₄ (Alexandrite)	0.23	0.25*	6.30	45.99	59.5	6	35.70	104.46
Al₂O₃ (Sapphire)	0.42	0.25*	5.0	35.23	178.8	5.5	98.34	287.75
Li Y F₄ (YLF)	0.06	0.25*	10.0	7.65	58.8	0.34	2.0	5.85
GSGG								
GGG	0.10		9.5					
Glasses								
LHG-5 (Phosphate)	0.00732	0.24	8.0	7.26	9.6	0.356	0.342	1.00
ED-2 (Silicate)	0.01254	0.24	7.7	9.62	12.9	1.00	1.290	3.77
Q-100 (Phosphate)	0.00820	0.25*	9.6	7.15	9.0	0.500	0.585	1.71
SiO₂ (Fused Silica)	0.00982	0.17	0.5	7.41	220.0	0.650	11.00	32.19

$$R_M = \sigma_M M_s, \qquad (2)$$

where σ_m is the ultimate tensile material strength and M_s a material constant given by (6)

$$M_s = \frac{K(1-\nu)}{\alpha E} \qquad (3)$$

Here K, ν, α, and E are thermal conductivity, Poisson's ratio, thermal expansion coefficient, and Young's modulus, respectively. Values of these parameters, M_s, σ_m, R_m, and a normalized R_M (relative to LHG-5 laser glass) are shown in Fig. 6. While some data is available for LHG-5, YAG, Alexandrite, Sapphire, and other materials, the physical properties of GGG and the co-doped material GSGG were not available. It is interesting, however, that according to Fig. 6, the average power capability of Nd:YAG is nearly thirty times that of LHG-5 while Alexandrite is over one hundred times, and that of Sapphire nearly three hundred times. While this comparison does not take into account other factors, for example, that the saturation fluence required to efficiently extract the optical energy, which is very large in Alexandrite, it nevertheless is a useful one in identifying promising HAP materials. It should be noted that crystalline materials have a much higher figure of merit than amphorous ones, principally because of a larger thermal conductivity and strength. In Fig. 7 we have shown a plot of the normalized maximum high average power for five optical materials as a function of pump surface area (for unity extraction efficiency). In order to obtain HAP from such materials, it is necessary to grow crystalline materials in much larger sizes than presently available. Nd:Glass for example, is presently available in pump areas up to 3×10^3 cm². For Nd:YAG, however, the largest pump areas available are only $\simeq 50$ cm², and it is not difficult to estimate that such sizes are capable of only a few kilowatts of HAP. Thus, to construct large crystalline HAP systems will require a substantial development program to grow significantly larger material sizes with good optical quality, or a clever implementation of segmentation technology. Optimization of Nd:Glass materials parameters, laser glass strengthening and prestressing may all lead to significant improvements in the power handling capability of amorphous materials.

Finally, in Fig. 8, we list a number of known limitations to the scaling of HAP laser devices. The importance of all of these effects in the design of HAP laser devices has been discussed previously (1,7), and place severe limits on the scaling rules presented here.

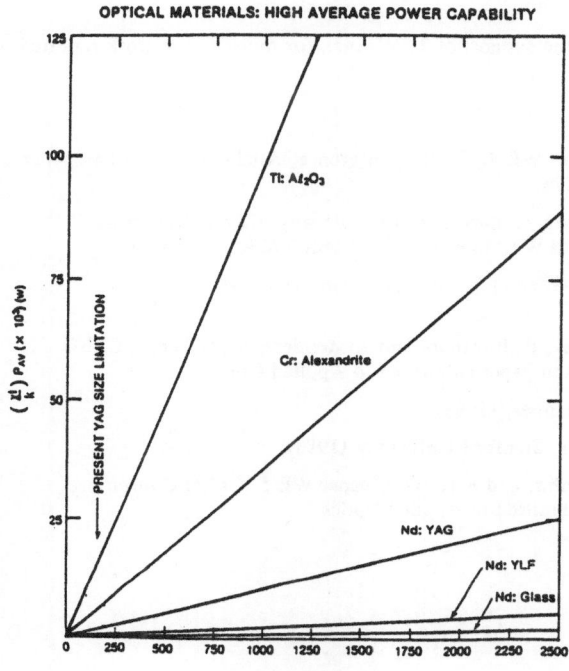

FRACTURE LIMITED SCALING LAWS

ROD: $P_{AV}{}^M = \dfrac{8\pi}{\sqrt{2}} \left(\dfrac{\eta_{ex}}{\chi} \right) R_M \, \ell_p$

CONVENTIONAL SLAB: $P_{AV}{}^M = \dfrac{12}{\sqrt{2}} \left(\dfrac{\eta_{ex}}{\chi} \right) R_M \left(\dfrac{w\ell_p}{t} \right)$

SANDWICH SLAB: $P_{AV}{}^M = \dfrac{24}{\sqrt{2}} \left(\dfrac{\eta_{ex}}{\chi} \right) R_M \left(\dfrac{w\ell_p}{t} \right)$

ACTIVE-MIRROR: $P_{AV}{}^M = \dfrac{6\pi}{\sqrt{2}} \left(\dfrac{\eta_{ex}}{\chi} \right) R_M \left(\dfrac{D^2}{t} \right)$

DISK-AMPLIFIER: $P_{AV}{}^M = \dfrac{3\pi}{\sqrt{2}} \left(\dfrac{\eta_{ex}}{\chi} \right) R_M \left(\dfrac{D^2}{t} \right) \sqrt{\eta_o{}^2 + 1}$

GENERAL: $P_{AV}{}^M = K \left(\dfrac{\eta_{ex}}{\chi} \right) R_M \left(\dfrac{V}{t^2} \right)$

$R_M = \sigma_M \, M_S$: MATERIAL FIGURE OF MERIT

$M_S = \dfrac{k \, (1 - \nu)}{\alpha E}$: MATERIAL CONSTANT   Fig. 8

ACKNOWLEDGMENT

The author gratefully acknowledges the support of TRW, Inc., for much of the work reported in this article.

REFERENCES

(1) D.C. Brown and K.K. Lee, paper WE 4, CLEO Conference, Anaheim (1984) and paper to be submitted to Applied Optics.

(2) W.S. Martin, Face-Pumped Laser, General Electric Company, CR&D Report 68-C-285 (1968), and J.C. Almasi and W.S. Martin, U.S. Patent 3,679,996 (1972).

(3) K. Tomiyasu and J.C. Almasi, General Electric Company, U.S. Patent 3,500,231 (1970).

(4) D.C. Brown, K.K. Lee, J. Kuper, R. Bowman, and J. Menders, paper WE 3, CLEO Conference, Anaheim (1984) and paper submitted to Applied Optics.

(5) W.B. Jones, Laser Focus, September, (1983).

(6) J.M. Eggleston, Doctoral Thesis, Stanford University (1983).

(7) D.C. Brown, K.K. Lee, K.J. Kuhn, and R.L. Byer, paper WE 5, CLEO Conference, Anaheim (1984), and paper submitted to Applied Optics

Modeling Studies and Experimental Performance of Slab Geometry Lasers

T.J. Kane and R.L. Byer

Ginzton Laboratory, Stanford University, Stanford, CA 94305, USA

ABSTRACT:
We report on progress made in the modeling of thermal effects in zig-zag slab geometry solid state lasers. A numerical model has been completed and verified. An analytic model exists which is adequate in many simple cases. These models predict that a laser using a slab of square cross-section could have low thermal focusing and birefringence.

A conventionally designed solid state laser contains a cylindrical rod of the active laser material. For average powers below a certain level, the rod design is optimal, as it is simple and well matched to the round laser beams that are usually desired. Above a certain level of average power, which varies from laser material to laser material, undesirable thermal effects distort the laser beam to an unacceptable degree. These thermal effects are well understood and can be estimated using the equations found in Koechner[1] and other sources. The zig-zag slab solid state laser design can greatly reduce these thermal effects. In this design, invented by General Electric researchers in the early seventies,[2] the active laser medium consists of a rectilinear slab, with optically finished surfaces facing the pump lamps. Confined by total internal reflection at the pumped faces, the laser beam zig-zags through the slab. If the slab is assumed to be infinite and to be uniformly pumped and cooled, then all undesirable optical effects are found to cancel, and the average power of the slab laser is limited only by the fracture of the slab.

In our talk we describe our work to develop a model of thermal effects which is more realistic than the "infinite slab" model. From the beginning, slab laser researchers have realized that finite size and non-uniform pumping and cooling would result in undesirable optical effects. We now have a quantitative model of these effects. Its numerical form gives detailed information for a general case slab, while its analytic form allows critical slab performance values to be estimated in many important special cases.

The numerical model requires as inputs the slab material properties and dimensions, the pumping distribution and the surface cooling coefficients. It then solves partial differential equations to find the stress and temperature distribution within the pumped slab. These partial differential equations are given in our earlier work.[3] These distributions are used to find the index of refraction distribution in the slab and to calculate the thermally distorted shape of the slab. Optical rays are then traced through the slab and a Jones matrix is found for each point on the slab aperture. These Jones matrices are used to find the thermal focusing and depolarization of the slab. Thus the three key undesirable thermal effects - focusing, depolarization and stress - are known and can be displayed graphically for the slab designer.

We have measured focusing and depolarization in glass and crystalline slabs and compared the measurements to the model. The depolarization has been predicted accurately for both large glass slabs and small Nd:YAG slabs, while the focusing predicted by the model is close to reality only for the small crystalline slab. The reason for the disagreement in the glass case will be discussed below.

For many special cases, the thermal effects can be accurately estimated analytically. The number of cases which can be treated is large, so only one will be presented here. It is the case of a slab which is uniformly pumped, cooled only at the pumped surfaces, and which has a square cross-section. For this case, we find the depolarization averaged over the slab aperture, the peak depolarization and the thermal focusing.

At small levels of depolarization, the depolarization of a zig-zag slab varies inversely with the square of the number of zig-zags in the slab, N , if the total heat load is held constant. The reason for this is not obvious, but is due to the reflection symmetry of the stress distribution about a plane which besects the slab and is perpendicular to the plane of the zig-zag path. This symmetry leads to a cancellation which results in the depolarization after all N zig-zags being no greater than after the first. Though the form of the relation between depolarization and slab material properties can be derived, the coefficient was found using the numerical model. Peak depolarization in a uniformly pumped slab of square cross-section can be estimated using the equation

$$D = 6\% \ (CP/N)^2 \tag{1}$$

where P is the total heat power dissipated by the slab (usually about 5% of lamp power). The parameter C is a material property given by

$$C = \frac{B}{4\lambda M_s} \quad \text{with} \tag{2}$$

$$M_s = \frac{(1 - \nu) \ k}{\alpha E} \tag{3}$$

and with α the coefficient of thermal expansion, E Young's modulus, B the stress optical coefficient, k the thermal conductivity, ν Poisson's ratio and λ the optical wavelength. The average depolarization is given by an equation with the same form as Eq.(1), but with a coefficient of 0.75%. Equation (1) agrees with experiment and with the numerical model to within a factor of two over a wide range of slab operating parameters. For comparison, the peak depolarization in a rod is given by Eq.(1) with $N = 1$ and the coefficient equal to 25% .

When a slab is uniformly pumped and carefully insulated to create a thermal distribution which is identical to that of the infinite slab, then focusing is due only to stress effects. Stress not only changes the index of refraction through the stress optical effect, but also leads to distortion of the reflecting surface. Both of these

effects cause focusing. This focusing is cylindrical and depends on
the polarization of the laser. As with depolarization, the form of
the equation for focal length is derivable, and only the coefficient
is found numerically. The thermal focal length of a pumped, square
laser slab, insulated on the unpumped surfaces, is given by

$$f = -3.53 \, t^2 \, M_s \, \cos(\theta) / B_{slab} \, P \tag{4}$$

where t is the thickness of the slab M_s and P are as before, θ
is the angle between the slab surface and the totally internally
reflected laser beam, and B_{slab} is, for the P polarization given by

$$B_{slab} = B_{=} \, [\cos^2(\theta) + \nu \sin^2(\theta)] + B_b [\sin^2(\theta) + \nu \cos^2(\theta)]$$

$$+ \frac{(1 - \nu^2) \, n \, \sin^2(\theta)}{E} \tag{5}$$

where $B_{=}$ is the parallel stress optical coefficient, B_b is the
perpendicular stress optical coefficient and n is the index of
refraction. Other variables have been defined previously. For the
S polarization the value of B_{slab} is given by

$$B_{slab} = B_b (1 + \nu) + \frac{(1 - \nu^2) \, n \, \sin^2(\theta)}{E} \tag{6}$$

For the s polarization, with Nd:YAG, the coefficient B_{slab} can
be exactly zero for $\theta = 15.2°$. Even for the P polarization, which
is preferred since Fresnel losses can be minimized near Brewster's
angle, the focusing is found to be far less than is the case for a rod.
     Our model is useful for the design of low aspect ratio slabs
because it fully explains edge effects, that is, effects due to the
finite extent of the slab perpendicular to its long axis. For small
aperture slabs, these edge effects are the primary source of thermal
distortion. For broad slabs, these edge effects are negligible over
most of the slab aperture and instead it is pumping and cooling non-
uniformity that limits slab performance. Our model does not calculate
the pumping distribution or the cooling coefficients, but rather uses
them as inputs. Unfortunately, these values are generally difficult
to obtain. This, we believe, is why our model does not accurately
predict focusing for the case of the large glass slab. The key to
good performance in large glass slab lasers is uniform and stable
pumping and cooling, and this model does not suggest ways to achieve
this. Nevertheless, it is now possible to estimate stress, depolarization and focusing if the thermal boundary conditions are known, and
this may prove useful in specifying the degree of pumping and cooling
uniformity required.

REFERENCES:
1.   W. Koechner, <u>Solid State Laser Engineering</u>, (Springer Series in Optical Sci.), New York, Springer-Verlag, 1976.
2.   W.S. Martin and J.P. Chernoch, "Multiple Internal Reflection Face Pumped Laser", U.S. Patent #3,633,126. 1972.
3.   J.M. Eggleston, T.J. Kane, K. Kuhn, J. Unternahrer and R.L. Byer, "The Slab Geometry Laser - Part I : Theory", IEEE J. Quant. Electr. vol. <u>QE-20</u>, p.289, (1984).

Slab Laser Development at MSNW: The Gemini and Centurion Systems

J.M. Eggleston and G.F. Albrecht

Mathematical Sciences Northwest, Inc., Bellevue, WA 98004, USA

Zig-zag optical path slab geometry[1] solid state lasers have lower focusing and birefringence than equivalent rod geometry lasers. By using the slab geometry, solid state systems can be scaled to higher repetition rates without sacrificing beam quality. At MSNW, we are investigating both Nd:YAG and Nd:Glass slab lasers, to understand their scaling and develop more efficient, high pulse rate solid state lasers.

The Nd:Glass laser under construction at MSNW, "Gemini", uses a novel pump geometry, shown in Fig. 1. In this design we place the flashlamps between two slabs in the same laser head. This is the opposite of the standard configuration, where a single slab is sandwiched between lamps. In the dual slab, sandwiched lamp configuration, light emitted from the lamps has a greater chance of being directly incident on the slabs without reflections from the large, usually inefficient, reflector structures that are typical of the standard configuration. Furthermore, light which passes through the slabs once, is reflected back through the slabs for a second pass, therefore achieving longer pump light absorption paths and greater coupling efficiencies between the lamps and the slabs.

The performance goal of the Gemini system is 10 J at 10 Hz in a Q-switched, diffraction limited mode. The system architecture is

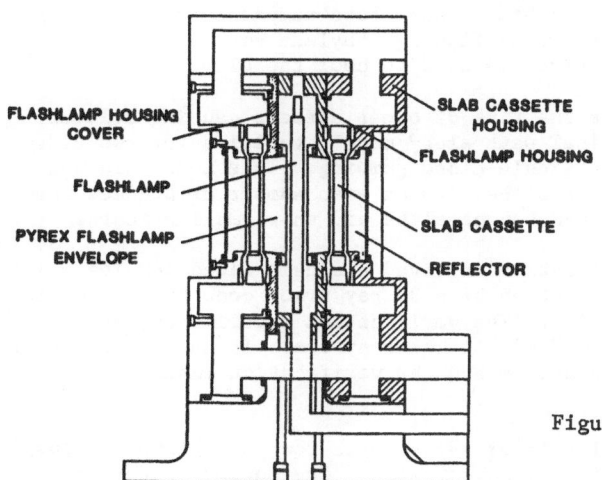

FLASHLAMP HOUSING COVER

FLASHLAMP

PYREX FLASHLAMP ENVELOPE

SLAB CASSETTE HOUSING

FLASHLAMP HOUSING

SLAB CASSETTE

REFLECTOR

Figure 1. Cross Section of Slab Glass Laser Head

a MOPA chain. A rod geometry, Nd:YLF oscillator will be used to achieve short pulses (<10 nsec), without optical damage problems. The glass slabs will amplify the oscillator output. The first of the Gemini slabs will be used in a triple pass arrangement, the second slab will be used as a single pass, final amplifier. The slabs are 6.3 mm x 5.6 cm x 35 cm in size and are composed of Hoya[2] 3.3 weight percent doped LHG5 glass. The central 30 cm of each slab will store approximately 25 J of optical energy at the time of extraction.

The device is nearing completion of the construction phase. System shakedown will begin shortly. Initial testing of all of the major subsystems has already been completed. Key aspects of the design geometry have been validated on a smaller pulse energy YAG laser, as well as thoroughly modeled.

In the proposed sandwiched flashlamp geometry, the slabs are pumped directly on one face and indirectly on the other. Thus the thermal deposition in the slabs is asymmetric. In this paper, this type of pumping is referred to as single-sided pumping. The major issues in single-sided pumping of a zig-zag optical path slab are: 1) creation of non-averaged thermal-optical effects, 2) pumping and cooling uniformity, 3) efficiency, 4) mechanical implementation. Both modeling and verification experiments were performed to minimize the risk that undesired effects would appear. These efforts are reviewed below.

Single-sided pumping will result in a bowing of the slab unless it is properly cooled. This bowing would directly distort an optical beam reflecting from its surfaces and cause unwanted focusing effects. Analytic analysis shows that to avoid the bowing, the stress on the two slab faces must be equal, which requires that both faces be at the same temperature. If a thermally conductive liquid coolant is in direct contact with glass slabs and the coolant temperature rise in traversing the slab is small, then this criterion will be met and bowing will not be a problem. If a gas cooling scheme is used and the temperature rise is significant, then careful thermal engineering is required to achieve equal face temperatures. In the Gemini system, a fast transverse flowing ethylene glycol-water mix, with a small temperature rise, is used to cool the slab.

In order to reduce the risk of other effects causing deleterious focus, a zig-zag optical path slab[3] model, verified for two-sided pumping, was modified for single-sided pumping. This model did not predict any new effects for slabs with an even number of bounces when the cooling was proper. Odd bounce slabs did show small effects.

Experimental verification involved a single-sided pumping demonstration, and an evaluation of a 3D ray trace code used to design uniform reflector geometries. The verification experiments, were performed on the "Centurion" Nd:YAG laser system. In the remainder of this paper, the Centurion system and the verification experiments will be discussed.

The "Centurion" slab laser system uses four flashlamps to pump a single 6 mm x 2 cm x 15.5 cm Nd:YAG slab. The maximum repetition

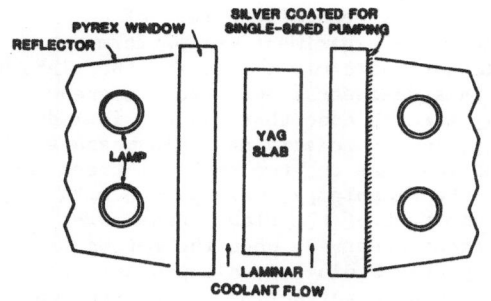

Figure 2. Centurion
Reflector Housing

rate is 100 Hz, and the laser has generated 70 W at full power. The
output is power supply limited and not limited by thermal effects.
The reflector structure is shown in Fig. 2. Pyrex windows separate
the transverse slab coolant flow from the longitudinal flashlamp flow.
The reflectors are gold-coated brass pieces, which have planar
surfaces at precise angles.

Figure 2 also shows where a silver coating was applied to the
back side of one of the pyrex pieces for the single-sided pumping
test. In this test, the oscillator performance and defocusing of a
single-pass beam, as a function of repetition rate, were measured and
compared to the equivalent results for two-sided (standard) pumping.
The single-sided pump output was 30 percent more efficient, than the
standard configuration. This improvement arises from the longer
absorption path achieved by double passing the pump light through the
slab. The results of the defocusing measurements are shown in Fig. 3.
The defocusing, at zero power, arises from the finite optical quality
of the Nd:YAG material. An improved optical quality slab is currently
being manufactured. Beyond the zero power effects, the ratio of defocusing strength (diopters) to pump power is constant and approximately equal for both single-sided and two-sided pumping. The
equality of the two terms cannot be exactly determined, as the ratio
of thermal power to optical power in the slab is not the same in the
two arrangements. However, it is clear that there are no major
defocusing effects from single-sided pumping.

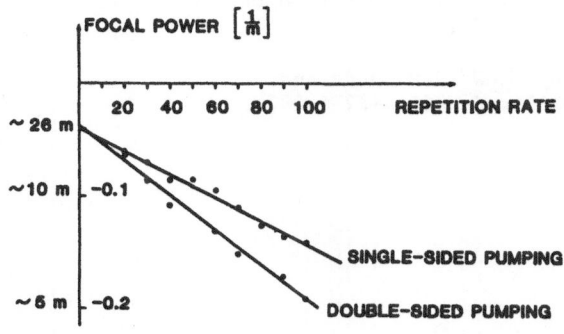

Figure 3. Experimental Demonstration of Single-Sided Pumping

The previous measurement was not as sensitive a test of de-focusing as it could have been, since the experiment was performed with a reflector structure that did not uniformly illuminate the slab. The Centurion reflectors, used in this demonstration, were designed with a 2-dimensional, geometric ray tracing code that did not include lamp refraction. Schlieren imaging, interferometry and gain measurements on the operational device, showed that the two small sections of the slabs, directly adjacent to the flashlamps, were pumped with 10 to 15 percent less energy than the rest of the slab. Pump power dips were predicted, identical to those observed, when the reflector structure was evaluated using an improved 3 dimensional, geometric, ray tracing code, "Trace3D", which includes refraction and absorption in the lamps. This new model was used to generate an improved reflector structure, which has since demonstrated gain uniformity to ±2 percent over the entire slab structure.

The Centurion reflector work verified the predictive capability of the Trace3D code. This code was used to design the minimal, but important, reflector structure currently being implemented on the Gemini system. This should result in good pump uniformity in the non-averaged direction, which is critical to achieving the potential of slab geometry lasers.

In conclusion, "Gemini", a 100 W, sandwiched flashlamp, Nd:Glass laser system has been assembled and should begin initial operation soon. Verification experiments, for the sandwiched flashlamp geometry, were performed on "Centurion", a nominal 70 W slab Nd:YAG laser. These experiments involved verification of a thermal-optical model, under single-sided pumping gometry and verification of Trace3D, a 3 dimensional, geometric flashlamp-slab coupling code.

REFERENCES

1. W.S. Martin and J.P. Chernoch, "Multiple Internal Reflection Face Pumped Laser," U.S. Patent #3,633,126 (1972).

2. Hoya Optics, Inc., 3400 Edison Way, Fremont, CA.

3. J.M. Eggleston, T. Kane, J. Unternahrer, and R.L. Byer, "Slab-geometry Nd:glass laser performance studies," Opt. Lett. 7, (9) pg. 405–407 (September 1982).

4. This work was supported by MSNW IR&D funds and in part by Contracts DE-AC06-83ER53157 and N00014-82-C-0802.

Part VII

Progress in Garnet
Host Solid State Lasers

Advanced Crystalline Laser Materials*

W.F. Krupke

Laser Program, University of California, Lawrence Livermore National Laboratory, P.O. Box 5508, Livermore, CA 94550, USA

Lawrence Livermore National Laboratory (LLNL) has recently undertaken an effort to develop new solid state lasers capable of operating at considerably higher performance levels (efficiency, average power, wavelength diversity, etc.) than attainable with current technologies. To guide this effort, we have performed analyses to establish the fundamental limits on the performance levels of solid state lasers in terms of known laser physics and design principles (gain, gain saturation, amplified spontaneous emission, parasitics, nonlinear self-focusing, etc.) and the physical properties of crystalline and amorphous materials.[1,2] A technical summary of these results, and the salient conclusions drawn from them, was presented by John Emmett at this conference. To realize the substantially higher performance levels permitted within these fundamental limits, we conclude: 1) that thin-plate geometries (zig-zag[3] or flow-cooled disk[4]) must be adopted and fully exploited through innovative engineering design solutions; 2) that efficiencies can be greatly improved by developing new types of efficient, high-irradiance pump sources and especially by instituting measures that rigorously maximize the utilization of the pump radiation generated by presently available efficient flashlamps; and 3) that new solid state laser gain materials must be developed which possess laser properties (stimulated emission cross-section, energy storage lifetime, saturation fluence, etc.) and bulk properties (mechanical, thermal, and optical) dictated by the applications of interest. To guide the laser materials research and development effort, we have analyzed the scaling laws for zig-zag and flow-cooled disk amplifiers, defined materials figures of merit, and developed laser performance applications maps.[2] Within this technical framework, we have embarked on an effort to synthesize, characterize, select, and develop new materials useful for a variety of robust laser applications.

To illustrate the leverage of identifying new laser gain materials responsive to operational requirements, consider the problem of developing an efficient, long-lived multijoule/pulse tunable blue laser. One generic approach is based on the use of a suitable Cr^{3+} or Ti^{3+} doped crystal in a flashlamp-pumped laser, efficiently generating multijoule pulses across the 800–1000 nm region, with a pulse duration sufficiently short to enable efficient harmonic generation in the blue. The technical properties of a gain medium capable of this service are highly coupled to one another. One logical string on

---

*Work performed under the auspices of the U.S. Department of Energy by Lawrence Livermore National Laboratory under Contract W-7405-Eng-48.

technical imperatives is as follows: to be long lived, the pump flashlamp must be operated at a loading small compared to the explosion limit; to provide a significant amount of pump energy at low load, the flashlamp pulse duration cannot be very short; to be efficient, the gain medium needs to couple well to the pump spectrum and to store energy for a time long compared to the pump pulse duration; however, the energy storage time, τ_f is inversely related to the stimulated emission cross-section, σ_L, the fluorescence (tuning) bandwidth, $\Delta\nu$, and the square of the index of refraction, n, according to the Fuchtbar-Ladenburg expression

$$\tau_f \leq \lambda_L^2/8\pi c n^2 \Delta\nu\sigma_L \tag{1}$$

where λ_L is the mean wavelength of the fluorescence emission band, and we have assumed that the fluorescence lifetime is equal to the radiative lifetime. Thus, long storage time and broad tunability drive σ_L to lower values. However, one cannot permit σ_L to fall below that value which results in adequate small signal gain and efficient energy extraction. For the latter, the output fluence of the laser must be set at one or two times the saturation fluence of the medium, $\Gamma_{sat} \simeq hc/\lambda_L\sigma_L$; and for long-lived operation, Γ_{sat} must be at least several times smaller than the single shot damage fluence, Γ_d.

How do present materials stack up against these requirements and what is the prospect for materials with significantly more favorable technical properties? Figure 1 shows graphs of Equation 1, assuming λ_L = 750 nm and $\Delta\nu$ = 1700 cm^{-1}, typical of chromium doped crystals.[5] Curves are drawn for n = 1.44 and n = 1.85, typical of fluoride and oxide crystals, respectively. Points for a number of chromium doped oxide[5,6] and fluoride[7,8,9] crystals are also shown in Figure 1 (it is assumed here, reasonably, that the radiative lifetime is equal to the fluorescence lifetime). Experience to date with flashlamps and optical damage suggests that τ_f values of at least tens of microseconds and Γ_{sat} values < 10 J/cm^2 are required. The index dependence of Equation 1 favors fluoride materials by roughly a factor of two in meeting these conflicting requirements (fluorides also appear to be

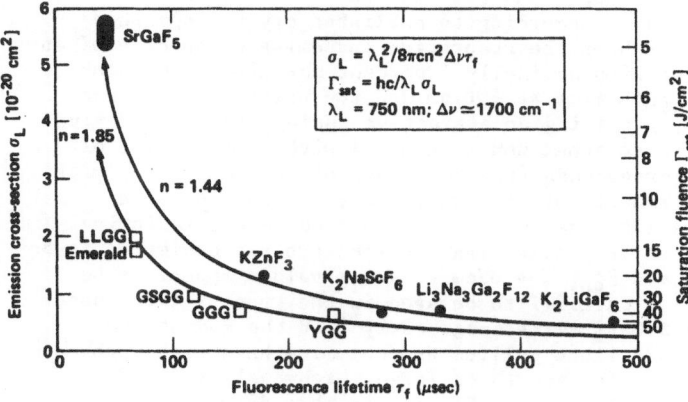

Figure 1. Parametrics of Cr^{3+} laser gain media.

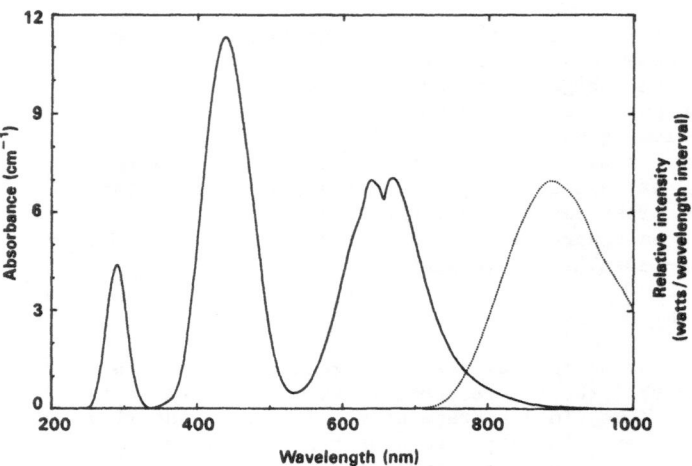

Figure 2. Absorption and emission spectra of
Cr:SrGaF$_5$. (Crystal grown by
H. Guggenheim, AT&T Bell Laboratories;
spectral data by M. Shinn, Lawrence
Livermore National Laboratory)

more immune to deleterious excited state absorption processes than
oxides[10]; at the same time, fluorides generally are less thermo-
mechanically robust than oxides and can sustain less thermal
loading[1]). In earlier work,[9] we examined the elpasolite (K_2LiGaF_6)
and the fluoride garnet ($Li_3Na_3Ga_2F_{12}$) possessing highly symmetric O_h
sites for the chromium substitution, which yield low σ_L transitions
and relatively high τ_f values. More attractive properties are found
for the perovskite[7,9] $KZnF_3$, presumably from a lowering of the site
symmetry from O_h due to charge compensation. Although these fluoride
materials may be quite useful for applications requiring laser pulses
of long duration (> μsec), for the present interest one needs to use
materials with yet more distorted sites for the chromium ion. The
tetragonal crystal[11] SrGaF$_5$ (and the aluminum[12] analog, as well)
offers this possibility, according to published crystallographic
studies.[13] Figure 2 shows the absorption and emission spectra of an
SrGaF$_5$ crystal doped with nominally 2 percent chromium. The peak
fluorescence wavelength lies at 890 nm and the half-width extends
from 800 to 980 nm, providing an attractive tuning range. The fluorescence lifetime is somewhat nonexponential with a mean value of
approximately 50 microseconds (the two types of gallium sites, with
slightly differing amounts of distortion, may account for this
behavior). If this lifetime indeed turns out to be the radiative
lifetime, then the corresponding laser parameters are estimated to be
$\sigma_L \sim 5 \times 10^{-20}$ cm^2 and $\Gamma_{sat} \overset{\sim}{\sim} 4$ J/cm^2. These values appear to be
quite promising for a material to be used in the type of laser under
discussion. Since SrGaF$_5$ melts congruently[11] at the modest temperature of 812°C, and since gallium and chromium have nearly the same
ion size, the prospect for growth of large, high-quality doped
crystals would appear promising. Efforts in this direction are
proceeding; at the same time, the average power handling capacity of

this crystal will likely prove to be less than desired for all applications, and we are continuing the search for additional fluoride and oxide crystals with more robust thermo-mechanical properties. Using the technical methodology illustrated by the example above, we are identifying and assessing novel solid state gain media for a variety of applications.

Acknowledgments

I am pleased to acknowledge and thank Dr. Howard Guggenheim of AT&T Bell Laboratories for growing the fluoride crystals discussed here, and Livermore colleague Dr. Michael Shinn, who performed the spectroscopic measurements on these crystals.

References

1. J. L. Emmett, W. F. Krupke, and J. B. Trenholme, Sov. J. Quantum Electron., 13, 1 (1983); see also UCRL-53344, Lawrence Livermore National Laboratory (November 1982).

2. J. L. Emmett, W. F. Krupke, and W. R. Sooy, UCRL-53571, Lawrence Livermore National Laboratory (October 1984).

3. W. S. Martin and J. P. Chernoch, U. S. Patent 3,633,126 (January 1972).

4. J. L. Emmett, J. F. Holzrichter, J. B. Trenholme and W. F. Krupke, "High Efficiency, High Repetition Rate Glass Lasers," Lawrence Livermore National Laboratory, Laser Program, Internal Memorandum (August 1977).

5. B. Struve and G. Huber, Appl. Phys., B30, 117 (1983).

6. M. L. Shand and S. T. Lai, IEEE J. Quantum Electron., QE-20, 105 (1984).

7. U. Brauch and U. Duerr, Optics Communications, 49, 61 (1984).

8. L. Andrews, Conference Record, Lasers '83, San Francisco, California (December 1983).

9. W. F. Krupke, Technical Digest, CLEO, Anaheim, California (June 1984).

10. L. Andrews, GTE Research Laboratory, Waltham, Massachusetts, private communication, 1984.

11. P. Julien and J. Chassaing, C. R. Acad. Sci., Paris, France, 271C, 139 (1970).

12. J. Ravez and D. Dumora, C. R. Acad. Sci., Paris, France, 269C, 331 (1969).

13. R. Von der Muhll, S. Andersson, and J. Galy, Acta Cryst., B27, 2345 (1971).

Use of Garnet Crystals as Laser Hosts

L.G. DeShazer

Hughes Research Laboratories, Malibu, CA 90265, USA

Garnets are a very numerous and versatile family of crystals, yet until recently only a few compositions (YIG, YAG, GGG) have been used extensively in lasers and opto-electronics. With the recent experience of Huber and Shcherbakov [1] in sensitizing Nd by co-doping with Cr in the garnet GSGG, greatly improving laser efficiency, interest has begun to develop to explore other garnets as hosts for new monovalent, divalent and trivalent laser-active ions for tunable solid state lasers and more efficient phosphors.

Systematization of crystallographic properties within the garnet class is extremely useful and demonstrates internal consistency of crystallographic data and the ability to predict the possible existence of new compounds. [2] The garnet structure is one of considerable chemical compliance with oxide sub-groups of silicates, aluminates, gallates, germanates, vanadates and arsenates with additional garnets such as fluoride garnets and hydrogarnets. We have been investigating the gallate and vanadate garnets, particularly $Gd_3Sc_2Al_3O_{12}$ and $NaCa_2Mg_2V_3O_{12}$.

We studied the spectroscopic and laser properties of Nd,Cr:GSAG where the crystal had 1.0 at% Nd and 0.75 at% Cr. We determined by spectroscopy that the laser cross section of Nd in GSAG is 2.2×10^{-19} cm^2, 40% of that of YAG. The free running laser oscillator performance of Nd, Cr:GSAG in the Hughes M-1 rangefinder system was compared to Nd:YAG. Q-switching of GSAG was achieved in both passive and active (electro-optic) modes, and the results are currently very good but preliminary. With these laser data, we determined that the Cr concentration should be increased to about double its present value for optimum laser performance. No solarization by flashlamp pumping was observed for GSAG which is an important advantage over GSGG. We are also exploring other garnet compositions such as $Gd_3In_2Ga_3O_{12}$ and $Gd_3Y_2Ga_3O_{12}$.

A garnet host of Cu^+, a potential blue-green laser ion, was investigated for the vanadate garnet $NaCa_2Mg_2V_3O_{12}$ where Cu^+ readily substitutes for the monovalent ion Na^+. The vanadate garnets are of particular interest because of their ease of preparation and the variety of chemical compositions possible. The fluorescence of monovalent Cu ions peaked near 530 nm and had a lifetime of 3.2 µs. We also successfully doped divalent Co into this garnet by substituting for Mg^{2+}. In addition, we are studying rare earth up-conversion phospors based on vanadate garnets such as $Na_2YMg_2V_3O_{12}$.

A study of color centers in YAG was undertaken for both undoped and Nd-doped crystals. We observed that the color center formation depends strongly on the growth facet direction, producing a "pinwheel" coloration of the boules when viewed along the <111> growth axis. We determined the quantum efficiency of the Nd dopant in the presence of these color centers. General spectroscopic features, including green fluorescence, was determined for the color centers believed to be associated with oxygen deficiencies.

References:
[1] D. Pruss et al., Appl. Phys. B28, 355 (1982).
[2] F. C. Hawthorne, J. Solid St. Chem. 37, 157 (1981).

Performance Study of a Cr:Nd:GSGG Laser

E. Reed and S. Guch, Jr.

GTE Government Systems, Mountain View, CA 94039, USA

Considerable effort has been expended in recent years toward improving the energy efficiency of neodymium lasers. Co-doping of Nd:YAG with Cr for this purpose has not been successful with flash-pumped lasers because of the longer Cr Nd transfer time (6.2 ms) [1]. By contrast, Nd in gadolinium scandium gallium garnet (GSGG) can be effectively sensitized by Cr because the transfer time is approximately 17 us [2]. The objectives of the work to be reported here were two-fold:

1. To maximize the efficiency of a flashlamp-pumped Cr:Nd:GSGG laser.

2. To fully characterize the material for its use in a flashlamp-pumped, Q-switched laser.

Table 1 summarizes the specifications and results of passive tests on the two Cr:Nd:GSGG rods, which were obtained from Union Carbide. The dopant concentrations were computed by Union Carbide based on known crystal growth parameters. The last four entries in the table were determined from measurements performed on one of the rods at Lawrence Livermore Laboratory [3].

Figure 1 shows the performance of a high efficiency, long-pulse laser using first a GSGG rod, and then a high-quality YAG rod. Resonator parameters are

Table 1   Cr:Nd:GSGG
laser rod specifications
and passive test results

- ROD SIZE	0.250 x 3.00 INCHES
- DOPANT CONCENTRATIONS (CALCULATED)	Nd 1.56 ATOMIC % Cr 1.18 ATOMIC %
- INSERTION LOSS (1064 nm)	4.5%
- OPTICAL QUALITY (1064 nm)	BETTER THAN ½ FRINGE FOR ROD LENGTH
- Nd FLUORESCENCE LIFETIME	256 MICROSECONDS
- Cr - Nd TRANSFER TIME	~15 MICROSECONDS

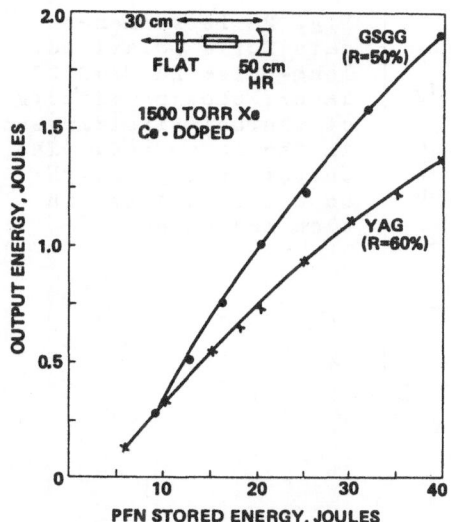

Fig. 1 Long-Pulse laser performance comparison for a Cr:Nd:GSGG rod and a Nd:YAG rod. The coupling mirror reflectivities are indicated

indicated in the Figure. This laser used a diffuse-reflecting, dry pump cavity. The laser wavelength was measured to be 10611.8 $\pm$0.1A.

The thermal focusing power (reciprocal of the focal length) of the GSGG rod was measured as a function of average flashlamp input power and compared to that of a YAG rod in the same laser. In this case the laser pump cavity was of standard design, with silver-plated reflectors and water cooling. The rod focal lengths were determined by measuring the laser's beam divergence, and then using an ABCD ray-matrix calculation for the resonator to find the value of the rod focal length which was compatible with it. The measurements show that, for a given input energy to the flashlamp, the thermal focusing power for the GSGG is 5.9 times that of YAG.

The thermal birefringence of GSGG is also more severe than that of YAG. Figure 2 shows the performance of a polarized GSGG laser. There is a precipitous drop in laser output energy, even at an average input energy of 100-200 watts.

Figure 3 shows the performance of a Q-switched GSGG laser, which used the standard silver-plated pump cavity. The electro-optic Q-switch was of a design which switched both polarizations, so that the rod birefringence did not contribute to a pump-dependent resonator loss. The laser performance data were fit with a mathematical model [4] in order to estimate the stimulated emission cross-section for Nd in GSGG. As shown, a value of 4.2×10^{-19} cm^2 best fits the data. It

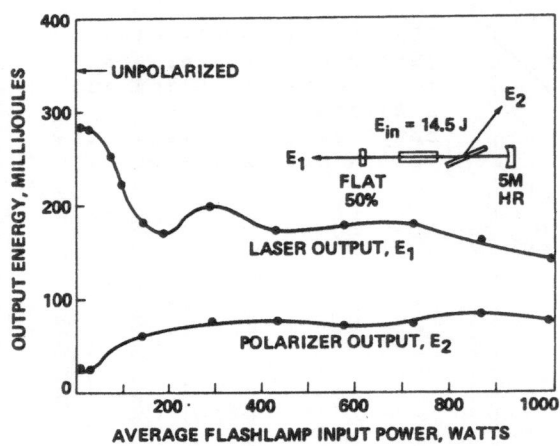

Fig. 2 Performance data for a polarized, long-pulse Nd:Cr:GSGG laser, showing effects of thermal birefringence in the laser rod. The output energy for the unpolarized laser is also indicated

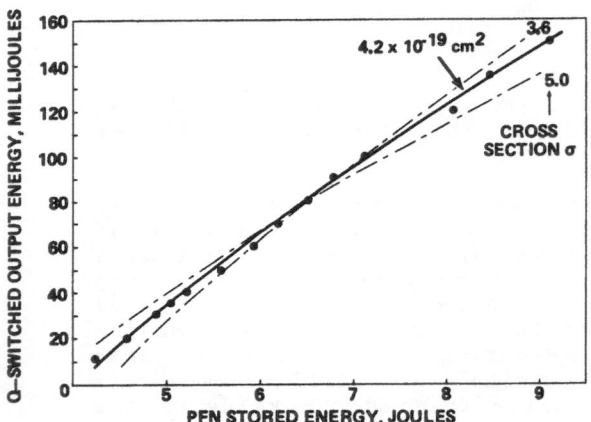

Fig. 3 Performance of the Q-switched Cr:Nd:GSGG laser, showing theoretical fits to the data for three different values of the stimulated emission cross section

is difficult to assign a meaningful uncertainty to this estimate; it is probably more revealing to compare this value to that obtained for Nd:YAG using the same data-fitting technique. The laser data for Nd:YAG indicated a cross section of 9×10^{-19} cm^2. Thus the laser data indicate that the cross-section for Nd in GSGG is about one-half that for Nd in YAG.

1. Z.J. Kiss and R.C. Duncan, "Cross-Pumped Cr^{3+} Nd^{3+}:YAG Laser System", Appl. Phys. Lett., vol. 5, pp. 200-202, 1964.

2. D. Pruss, G. Huber, and A. Beimowski, "Efficient Cr^{3+} Sensitized Nd^{3+}: GdScGa-Garnet Laser at 1.06 um", Appl. Phys. B28, pp. 355-358, 1982.

3. John Caird, Frank DeMarco, Mike Shinn, and Ray Wilder, Lawrence Livermore Laboratory.

4. R.H. Dishington, "Energy Extraction Optimization in Lasers", Proc. SPIE, vol. 69, pp. 135-147, 1975.

Part VIII

Growth of Lasers and
Nonlinear Materials

Growth of Codoped Garnet Crystals[†]

R. Uhrin and R.F. Belt

Airtron Division Litton Industries, Inc., 200 East Hanover Avenue, Morris Plains, NJ 07950, USA

Renewed interest in scandium substituted gadolinium gallium garnet (GSGG) has resulted from the fairly recent report of high slope efficiency in Cr^{3+} sensitized Nd^{3+} doped GSGG.[1] Investigators have strived to duplicate and improve upon the reported results of a twofold increase in slope efficiency over Nd:YAG at the same Nd concentrations. Concurrent growth efforts have also been made to obtain other suitable materials having the same lasing efficiency as codoped GSGG. These are sought because of the limited availability and very high price of the high purity scandium oxide component. These latter factors are of importance due to undeveloped resources in the U.S. and sole source availability from the U.S.S.R. Additionally, further purification of this material is required, probably due to inferior Soviet processing techniques or a desire to provide an oxide of only 3-9's purity.

To circumvent this scandium oxide difficulty Airtron has examined a codoped calcium, magnesium and zirconium substituted GGG while others have studied crystals such as the codoped aluminum analog of GSGG. The former undoped crystal has the advantage of growth in a large size with low bulk strain, but the first tests with the Cr,Nd doped material revealed color center formation during lasing. This behavior may be controllable and could lead to the availability of large plates for slab lasers if the performance characteristics are favorable. The latter crystal has the advantage of being able to be grown at the stoichiometric congruent melting composition but seems to possess a thermal conductivity similar to other aluminum garnets. This leads to faceting and a centralized core formation in the crystal and limits the potential size for large plates. At the same time efforts to eliminate the core formation lead to a degradation in crystal quality. A comparison of laser rod performance between this material and Nd:YAG has not been reported.

The initial growth efforts with codoped GSGG have been encouraging. Although an exact crystal composition and dopant concentration have not been formalized, the approach has been to obtain a crystal concentration of 1-2 x 10^{20} cm^{-3} for Cr and Nd. A slope efficiency of about 7% was achieved at concentrations of 0.9 x 10^{20} cm^{-3} and 1.6 x 10^{20} cm^{-3} for Cr and Nd respectively in a 5mm diameter x 75mm long laser rod. This compared favorably with a slope efficiency of 2.5% obtained with a Nd:YAG rod of the same size.[2] Even better performance is anticipated for higher Nd concentrations coupled with better Cr sensitization.

One unusual aspect of the GSGG crystal growth, however, is the occasional presence of an absorption band overlapping the 1.06μm laser output. The presence of this absorption does not appear to be systematic, but it is assumed to be associated with a Cr^{4+} ion arising from the lattice defect structure. Site selection in the garnets is highly dependent on ionic radius. The following table of radii, as calculated by Shannon and Prewitt[3], indicates the general selectivity of the various ions.

Table I

Calculated Ionic Size in the Garnet Lattice

Dodecahedral	Octahedral	Tetrahedral
Gd^{3+} – 1.06Å	Ga^{3+} – 0.620Å	Ga^{3+} – 0.47Å
Sc^{3+} – 0.87Å	Sc^{3+} – 0.730Å	Cr^{4+} – 0.44Å
	Cr^{3+} – 0.615Å	
	Gd^{3+} – --- *	

*Known to occupy the octahedral site as well

This indicates that an oxidized state of the Cr ion can exist in the tetrahedral site. Thus, depending on the defect structure, the optical absorption may be associated with Cr due to the oxidizing nature of the growth atomosphere. The exact nature of this defect is, of course, not understood at this time.

A more serious problem, however, may be the presence of the temperature induced growth striations which lead to localized index of refraction variations. Though objectionable, these are not a serious problem in a material such as Nd:YAG. However, the gallium garnets are commonly grown at rates that are an order of magnitude greater than that utilized for Nd:YAG. Consequently the extent of the growth striations is enhanced by temperature fluctuations arising from the diameter feedback control loop of the growth system.

Poor results have been achieved initially in the case of calcium, magnesium and zirconium substituted GGG crystal codoped with Cr and Nd. Lasing has been achieved at about 1% slope efficiency up to 100 Joule input where solarization occurs.[4] Under this condition the rod assumes a red color which bleaches back to green in about 24 hours. This mechanism may be related to the effect observed in the codoped GSGG and indicates very strongly an association with the lattice defect structure. It should be observed that charge compensation is required in this crystal with tetravalent zirconium on the octahedral site offsetting divalent calcium and magnesium on the dodecahedral and octahedral sites respectively. For research purposes crystals of nominal 25mm diameter have been grown, but crystals of 75mm diameter could effectively be produced where required.

†Portions of this work were supported by the M.I.T. Lincoln Laboratory under Purchase Order AX-31087 and the D.O.E. under Contract DE-AC08-84DP-40198.

References

(1) D. Pruss, et al., Appl. Phys. B, <u>28</u>, 355 (1982).

(2) P. Moulton, Private Communication.

(3) R. Shannon and C.T. Prewitt, Acta Cryst. B, <u>25</u>, 925 (1969).

(4) K.C. Liu, Private Communication.

Growth of KTP*

G. Gashurov and R.F. Belt

Airtron Division of Litton Industries, Inc., 200 East Hanover Ave., Morris Plains, NJ 07950, USA

KTP, potassium titanyl phosphate, is one of the best and most recent crystals to be developed as an optical second harmonic generator. The growth of mm single crystals of KTP and its characterization were carried out 10 years ago by a group of E.I. DuPont scientists.[1] Airtron has worked on the growth for the last 6 years and a characterization study of KTP was made by Y.S. Liu of General Electric 2 years ago.[2]

$KTiOPO_4$ belongs to the orthorhombic point group mm2 (space group $Pna2_1$) which lacks a center of symmetry. KTP is transparent over a large wavelength range, chemically and thermally stable, and nonhygroscopic. It has high nonlinear coefficients, large angular and temperature bandwidths and a high damage threshold. All these properties make KTP an outstanding frequency doubler. This paper gives a brief description of a study by Airtron which resulted in the development of a relatively simple, economically feasible and reproducible growth technique of KTP crystals.

The incongruent melting point of KTP precludes a direct growth method such as the Czochralski. Ordinary flux methods also do not appear promising because of temperature control, seeding, size, morphology, and generally poor quality. The technique developed by Airtron involves the hydrothermal growth of KTP at 600°C and 25,000 psi. The growth is carried out in gold lined autoclaves made of special alloy with modified Bridgman seals. The autoclave dimensions are 3 inch O.D., 1.5 inch I.D. and 21 inch in length. A typical gold liner contains a seed rack welded to a perforated disc on the bottom. The presence of a disc makes it possible to reduce the mass transfer rate and thus helps to separate the liner into two nearly isothermal regions, one for dissolving and the other for growing.

In a typical run KTP nutrient is placed in the bottom or dissolving part of the liner, seeds at the top, flux and water are added, and the liner is then welded shut and inserted into the autoclave. Pure water is added to the volume between liner and autoclave. The autoclave is then sealed and the system brought to operating temperature. The desired temperature gradients are established and pressure is adjusted by pressuring or bleeding the system. The run duration is usually between five and six weeks.

The seeds are KTP crystal slices about 1mm thick with essentially parallel (011) faces. Typical growth rate per side on (011) ranges between 1mm/week and 1.5mm/week. With these rates a growth run of six

weeks produces crystals sufficiently large to yield KTP cubes measuring
up to 5mm. The maximum size is limited primarily by the size of the
growth autoclave which is now 1.5 inch diameter. The presence of the
seed and its misorientation or defects precludes the utilization of the
entire crystal; thus, only both halves of the crystals are sources of
useful material. Airtron is undertaking a major development program
sponsored by the Air Force to grow larger crystals from 3 inch I.D.
autoclaves. This system should permit production of KTP cubes measuring up to 1cm. The larger cubes should permit higher power operation
in many Nd:YAG systems.

The primary application of KTP has been directed to external cavity placement in a Q-switched Nd:YAG laser. A maximum energy conversion efficiency of 60% was observed for a 15ns pulse at an incident
power density of 250 MW/cm^2. The bulk damage resistance of KTP is
approximately 300 MW/cm^2. With a conversion efficiency of 30-60%, KTP
can yield a few watts average power at 532nm. In addition to the 1.064
μm output, the 1.30 and 0.94μm lines of Nd:YAG can be also doubled by
KTP.

*A portion of this work was sponsored by the Air Force, Contract No.
F33615-78-C-1523.

REFERENCES

1.   F. C. Zumsteg, J. D. Bierlein, and T. E. Gier, J. Appl. Phys. 47,
     4980 (1976).

2.   R. F. Belt, L. E. Drafall, G. Gashurov, and Y. S. Liu, Interim
     Report on Phase I, Contract No. F33615-78-C-1523, Airtron Division
     of Litton Industries, Inc., Morris Plains, N. J., February, 1984.

Growth of New Laser Crystals

M. Kokta

Union Carbide Corp., 750 S., 32nd Street, Washougal, WA 98671, USA

Introduction

The main objective in crystal growth is the high crystal quality; and it depends on approaching closely the quilibrium between the solid and liquid phases. The actual driving force for crystal growth is a deviation from equilibrium condition favoring the solidification. Such deviation, if maintained constant in magnitude, results in steady state growth process.

Another necessary condition for high crystal quality is congruency of melting and solidification of the compound. If both of the aboe conditions are satisfied, a suitable technique such as pulling may be applied. The pulling technique or its modifications offer important advantages to the crystal grower such as visibility, good fluid mixing, fluid flow control and ambient atmosphere control.

Crystal Composition Control

The compositional uniformity of the crystal is related to crystalline quality via lattice parameter variations. The congruency of melting does not guarantee the compositional uniformity, especially in crystals for application as lasers. In laser crystals, the active ion(s) (dopants) are usually of different species than the constituents of host lattice.

The dopant distribution is governed normally by $C_s = kc_0(1-g)^{k-1}$. Where C_s is dopant concentration in crystal, C_0 is starting dopant composition, g is ratio between the weight of crystal and original weight of melt, and k is the distribution coefficient. At $k = 1$, the material is truely melting congruently and crystal composition is maintained constant unless excessive evaporation of any of the constituents causes the material balance to break down. In any system where 'k' deviates from unity, the composition variations must be expected, and minimized by manipulating the value of g.

Another aspect of maintaining uniform composition is raw material purity and their handling prior to crystal growt. Handling includes drying, densifying, sintering or active atmosphere treatment.

Practical Examples

1. Growth of Gadolinium Scandium Aluminum Garnets. One of the candidates for tunable vibronic lasers is gadolinium scandium aluminum garnet (GSAG). This material could have some advantages compared to gadolinium scandium gallium garnet in its higher heat conductivity, and its resistance to color center

formation (solarization). GSAG was grown recently by Union Carbide doped with Cr^{3+} , as well as with $Nd^{3+}Cr^{3+}$ pair .

Crystals up to $4\frac{1}{2}$" long, 1.4" in diameter were grown by the Czochralski technique applying pull rates from 0.012"/hour to 0.06"/ hour and rotation rates from 6 to 20 rpm, under N_2 atmosphere. The crystals were grown with typical garnet morphology similar to YAG crystals. Distribution coefficient for Cr^{3+} is closely approaching unity. Chromium doping concentrations are limited in this material to about 2.2% atomic of octahedral sites.

2. Growth of Titanium doped Sapphire.
The growth of sapphire presents complications which are unexpected for such simple compounds. Defects are created by oxygen non-stoichiometry gases dissolved in the melt, as well as impurities present in raw materials. An attempt to incorporate sizeable amounts of Ti^{3+} in corundum structure magnify already existing difficulties . However, careful control of growth process via growth rates combined with ambient atmosphere control allows material with good quality to be grown.

Titanium Sapphire grown by UCC ranges from 1.5" in diameter to 2.5" in diameter, and up to 14" long. Growth rates applied are from 0.012"/hour to 0.1"/hour, under N_2 ambient atmosphere.

The scattering in the crystal as well as the broad band absorption in 750 nm to 1,000 nm is related to dopant concentration. The distribution coefficient $k = 0.11$ is lower than the k_{Nd} in YAG. The nature of the above broadband absorption remains to be explained.

Growth of Nd:YAG and Cr, Nd:GScGG
A Comparative Study

F.J. Bruni

Material Progress Corporation, 93 Stony Circle,
Santa Rosa, CA 95401, USA

I. INTRODUCTION

Nd:YAG is a well developed solid state laser material. It has been commercially available for about two decades and has withstood competitive pressures from a number of other materials to remain the basis of the largest single family of laser systems. YAG's dominance of solid state lasers stems from a combination of physical and chemical properties of the host material, yttrium aluminum garnet. Also, the growth and characterization of neodymium-doped YAG crystals has been refined over 20 years to the point where it is a fairly well understood system. The inevitable compromises required to grow this material are understood and acceptable within the commercial limitations of a mature product market.

Gadolinium scandium gallium garnet (GScGG), on the other hand, is a relative new comer to the laser community. Interest in this material stems less from the properties of the host per se than from the potential performance improvement brought about by co-doping with chromium. Although these two garnets have many similar properties their crystal growth parameters differ considerably. It is the differences in their physical and chemical properties that create the limitations that ultimately dictate the size and perfection of laser components available from these materials.

II. MATERIAL PROPERTIES

It is useful at the outset to list some of the physical and chemical properties of potential laser materials and compare their impact on laser operation versus crystal growth performance. Table I compares the relative importance of several physical properties of the laser host crystal to the laser engineer and the materials scientist.

The laser engineer desires a hard material for good optical finishes on the surfaces and resistance to damage but a high hardness correllates to a high melting point. The crystal grower would prefer a material with as low a melting point as possible to simplify crucible selection,

Table 1
Desired Properties of a Laser Host Crystal

Property	Laser Engineer	Crystal Grower
Hardness	High	Low
Thermal Conductivity	High	Very High
Crystal Structure	Cubic	Cubic
Chemical Stability	Moderate	Very High
Chemical Constituents	N.A.	Single Component
Dopant Segregation Coeff.	N.A.	Unity

heating method, etc. Similarly, the crystal growth engineer desires a material that is virtually unreactive chemically. While chemical stability is a concern to the laser optics engineer, it is a minor one because the systems are operated near room temperature in an environment that can be carefully controlled. A high thermal conductivity is important for both, growing and operating laser crystals; in the former case to transport the heat of fusion away from and to maintain the high temperature gradients needed to stabilize the growth interface and in the latter case to dissipate the waste heat of operation.

In the case of chemical constituents, the materials scientist desires a host system that behaves as a single component (i.e. freezes congruently). Sapphire and YAG satisfy this requirement and GScGG appears to as well. The effects of the dopant species on the congruency of solidification are set aside temporarily for this discussion regarding the host material. However, as·the last item in Table 1 points out, the segregaton or distribution coefficient of the dopant is of great concern to the crystal grower. Ideally a dopant that distributes itself uniformly between the solid and liquid phases is desired. The greater the deviation of the distribution coefficient from unity, the more complex becomes the task of the crystal growth engineer.

The interrelationships between these properties combine to determine the growth environment of the material as will be described below.

III. CRYSTAL GROWTH

These garnets are readily grown by the Czochralski method using an iridium crucible. Typically a nitrogen atmosphere containing a small amount of oxygen (800 ppm in the case of YAG and up to 3% in the case of the gallium containing garnets) is used at atmospheric pressure. In the case of the undoped crystals, growth rates are reasonably high for oxides, 8-10 mm/hr, which will yield an 8-10 inch long crystal in a day. The addition of dopants changes matters considerably, particularly for YAG.

The segregation coefficient of neodymium in YAG is 0.18, i.e. the ratio of the concentrations of neodymium in the solid and liquid at equilibrium is 0.18. This immediately becomes the single, dominant issue in the crystal growth of YAG. Growth rates, temperatue gradients and the shape of the crystal become dictated by the need to stabilizethe growth interface in the presence of the neodymium gradient ahead of it. The high temperature gradients required to stabilize the growth interface augment the material's natural tendency to grow with a solid-liquid boundary curved convexly toward the liquid. This combined with the formation of (211) type facets on the growth interface leads to the formation of a "core" of strained material along the crystal's physical axis.

In the case of the gallium containing garnets, gadolinium gallium garnet (GGG) and gadolinium scandium gallium garnet (GScGG), the distribution coefficient of neodymium is roughly 0.6 - 0.7 and that of chromium is near unity for GScGG but about 3 for GGG. These values make the growth of Cr,Nd:GScGG more tractable from the standpoint of dopant segregation. However, for these garnets the domi-nant material characteristic impacting the crystal growth is the chemical decomposition of Ga_2O_3 and subsequent reaction with the iridium crucible. It is this reaction that leads to the higher oxygen levels in the ambient atmosphere in order to stabilize Ga_2O_3. The higher oxygen concentration in turn leads to a requirement for lower temperature gradients in order to keep the iridium crucible from oxidizing. It is desireable to protect the crucible not so much from the economic considerations that arise from loosing the metal but more so from the fact that some fraction of the oxidized iridium ends up in the crystal in the form of metallic inclusions.

A beneficial effect of the lower temperatue gradient is the fact that it becomes more readily possible to grow the crystal with a near planar interface thus eliminating the central "core". This permits the preparation of samples using a much larger fraction of the crystal volume. How-ever, the lower gradients also lead to a lessened interface stability which is characterized by more intense growth striations.

In summary, whereas Nd:YAG is composed of very stable oxides and its crystal growth is dictated by the segregation coefficient of neodymium, it is the chemical stability of gallia that predominantly determines the growth environment of GScGG. While it may be of minor inportance for small laser systems (up to 10 cm³), the occurrence of small, iridium inclusions in the crystal may become the ultimate limitation on the preparation of really large Cr,Nd:GScGG samples (~200 cm³).

IV. LASER PROPERTIES

Because it is composed of more massive constituent ions, the thermal conductivity of GScGG tends to be lower than that of YAG. This limits its ability to compete with YAG in rod configurations but is less of a problem for slab geometries where the two dimensional nature of the thermal distortion does not have lensing effect on the beam.

At the present time there are two phenomena that occur periodically in Cr,Nd:GScGG crystals that limit their laser performance. These are a tendency to solarize or darken on exposure to UV radiation and an anomalous absorption band at the one micron wavelength. GGG has been known to solarize for some time and, while not unheard of, this phenomenon rarely occurs in YAG. The one micron absorption varies enough from crystal to crystal to suggest that it is impurity related or a result of some growth parameter such as oxygen level in the growth atmosphere rather than an intrinsic property of the material.

V. SUMMARY

To conclude, the physical and chemical properties of YAG and GScGG dictate considerably different growth conditions in order achieve single crystals with high optical quality. At the same time these different growth conditions lead to a final crystal that, fortuitously in the case of GScGG, is more readily fabricated into slab lasers. However, a greater understanding of the material science of GScGG is needed before it can be grown into laser crystals with the reproducibility of YAG.

Growth of Alexandrite and Garnet Laser Crystals

M.H. Randles, D.G. Dawes, and C.R. Perleberg

Allied Corporation, Synthetic Crystal Products,
Charlotte, NC 28231, USA

Alexandrite

Alexandrite is chrome-doped chrysoberyl, $Cr:Be\ Al_2O_4$. This crystal possesses excellent strength, hardness, thermal conductivity, and shock resistance, all requirements for a rugged laser host. The chrome substitution provides the characteristic vibronic transitions which generate laser output tunable from 700 to 815nm. The dopant substitution for the aluminum is varied in the range of 0.06 - 0.18 at %. Typical laser rods are 5 to 9.5mm diameter by 10-11cm long with lower doping levels specified for the larger diameters to optimize flashlamp pumping efficiency.

Alexandrite crystals up to 2 inches in diameter by 6 inches long have been grown by the Czochralski technique from iridium crucibles. An automatic diameter control system provides for smooth power control with very little operator intervention. As a precondition for Alexandrite production all beryllium containing dusts from raw material preparation,growth, and crystal fabrication are exhausted through absolute filters to ensure a safe working environment.

GSGG[1]

Gadolinium scandium gallium garnet (GSGG) has been grown under conditions similar to those practiced for GGG growth. Crystals 1½ inches in diameter by 4 inches long were pulled from a 3 x 3 inch iridium crucible with gallium suboxide formation being suppressed by a 2% oxygen atmosphere.

GSGG crystals were co-doped with Cr and Nd at nominal concentrations of 2×10^{20} cm^{-3}. The distribution coefficients of Cr and Nd were assumed to be 1.0 and 0.65[2]. Differential thermal analysis (DTA) performed at Allied's Corporate Research Center indicated a partial substitution of Gd on the octahedral site at approximately 0.08 in the garnet formula unit. This agrees with earlier work by Brandle and Barns[3]. With this assumption the search for the congruent composition in the co-doped system led to variations in the Sc-Ga ratio. Rather than DTA, boule growth with subsequent lattice constant measurements was chosen as a sensitive test for congruency. While not exactly congruent, the following composition expressed in formula units resulted in less than 1 milliangstrom difference from top to bottom of the boule:

$$Gd_{3.03}Nd_{.071}Sc_{1.57}Cr_{.05}Ga_{3.3}O_{12}$$

The quality of the Cr:Nd:GSGG boules appears very good. No iridium inclusions have been seen and neutron activation analysis verified an iridium content of 1-2ppm which compares with Nd:YAG and Alexandrite. The GSGG boules are grown with a flat interface and therefore show no "core" strain. Optical distortion is less than ½ fringe per inch at 1.15 micron. Optical losses at the lasing wavelength of 1.06 microns measure 0.001 to 0.005 per cm, again comparable with Nd:YAG and Alexandrite. GSGG behaves almost exactly like GGG even with respect to the occasional formation of dislocations during the transition from deep interface to flat interface growth.

The growth of GSGG is clearly in an early development stage. However, successful production scale-up of other garnet crystals has been demonstrated. At Synthetic Crystal Products Nd:YAG is routinely grown at 3-inch diameter and GGG at 4-inch diameter. Therefore, given the success of this early effort and the similarity with GGG it is almost certain that GSGG either singly doped with Cr or co-doped with Cr and Nd will be available in large sizes.

References

1. GSGG work sponsored in part by DOE contract DE-AC08-84DP40203 in coordination with Lawrence Livermore National Laboratory.

2. Per J. A. Caird at Lawrence Livermore National Laboratory.

3. C. D. Brandle and R. L. Barns, J. Crystal Growth 20, 1 (1973).

Part IX

Nonlinear Frequency
Conversion and Tunable Sources

Wavelength Conversion
via Stimulated Raman Scattering

D.A. Rockwell

Hughes Research Laboratories, 3011 Malibu Canyon Road,
Malibu, CA 90265, USA

In advanced remote sensing scenarios, the availability of tunable
solid state lasers offers a highly practical alternative to tunable
dye lasers, which represent the previous state-of-the-art.
Nevertheless, as with dye lasers, coverage of all the portions of the
visible and near IR spectrum of interest may not be complete.
Therefore, it is natural to consider methods of extending the laser
tunability range using nonlinear optical techniques. This paper
briefly reviews the status of devices exploiting the nonlinear process
of Stimulated Raman Scattering (SRS).   Other nonlinear techniques
are described elsewhere in this volume.

The Raman effect is an inelastic light scattering process.[1] In one
version, called a Stokes process, a photon interacts with a medium
(e.g. a pressurized gas), and leaves with less energy and a longer
wavelength than it initially had. In this case the medium is left in
an excited state. An anti-Stokes process may also occur in which a
medium initially in an excited state is left in the ground state. In
this case the scattered photon has more energy, and a shorter wavelength, than the incident photon. In both cases the wavelength shift
is a characteristic of the Raman medium. For many reasons, mostly
relating to efficiency and practicality, the anti-Stokes process has
not been exploited as a wavelength conversion technique. Hence, it
will not be discussed here.

Raman lasers have recently become the subject of intense developmental
efforts because of many outstanding features, including i) energy
conversion efficiencies >50%, ii) high spectral stability (equal to
that of the pump laser), iii) demonstrated energy scaling[2,3] to
>1 joule, and pulse repetition frequency scaling[4] to 60 Hz and beyond,
iv) relatively large wavelength shifts, v) passive, self-contained
configurations having demonstrated[4] long lifetimes (>10 million shots
at ~0.9 watts average output power), and vi) a large variety of
available Raman media. Using common gases such as hydrogen and methane
at pressures ranging from ~ 10 - 50 atmospheres, laboratory devices
have been scaled to average powers approaching 10 watts. Practical
engineering and producibility issues are also being addressed in the
MELIOS program.[5] Sponsored by the Army NVEOL, and awarded to the
Hughes Aircraft Company, the MELIOS is an eye-safe range finder
produced by Raman shifting the output of a Nd:YAG laser in pressurized
methane. This program represents the first attempt to put a reliable,
engineered Raman laser in the field.

Raman lasers can extend the tunability range of tunable solid state
lasers. For example, using a single Stokes process in methane
(frequency shift = 2917 cm^{-1}), the 1.6 - 2.1 μm, tunability range of

$CO:MgF_2$ can be translated to $3-5.4$ μm with a spectral "hole" in the region near 3.3 μm where methane shows IR absorption. Alternatively, using a single Stokes process in hydrogen (frequency shift = $4155 \, cm^{-1}$), the $700-800$ nm tunability range of Alexandrite can be translated to $\sim 990-1200$ nm. Many other possibilities exist. For example, it is possible to design a single Raman device such that the Raman shifted radiation acts like a new pump beam that is Raman shifted a second time[2,3] providing twice as much frequency shift. Also, by using gas mixtures, it is possible to achieve a net frequency shift that is the sum of the individual frequency shifts.[7] Finally, although most of the work on practical Raman wavelength conversion devices has involved vibrational energy levels in simple gas molecules, producing frequency shifts that range from $\sim 1000-4000 \, cm^{-1}$, shifts $\sim 100-500 \, cm^{-1}$ are possible using rotational energy levels,[8,9] or larger shifts $>10,000 cm^{-1}$ have been observed using electronic energy levels in metal vapors.

These Raman device capabilities are available with virtually any pump laser source which might be considered for remote sensing. As the maturity of these devices increases in the next few years, there is no doubt that they can play a major role in a wide variety of high power tunable laser applications.

REFERENCES:

1.   See, for example, G. Herzberg, Molecular Spectra and Molecular Structure, vol. I, Spectra of Diatomic Molecules (van Nostrand Reinhold, New York, 1950).
2.   H. Komine, E.A. Stappaerts, S.J. Brosnan and J.B. West, Appl. Phys. Letts. 40, 551 (1982).
3.   S.F. Fulghum, D.W. Trainor, C. Duzy and H.A. Hyman, IEEE J. Quant. Electr. QE-20, 218 (1984).
4.   D.G. Bruns, H.W. Bruesselbach, H.D. Stovall and D.A. Rockwell, IEEE J. Quant. Electr. QE-18, 1246 (1982).
5.   AN/PV-6 Mini Eye-safe Laser Infrared Observation Set (MELIOS), U.S. Army Contract #DAAK20-83-C-0171, Night Vision and Electro-Optics Laboratory, Ft. Belvoir, Virginia.
6.   G. Herzberg, Molecular Spectra and Molecular Structure, vol. II, Infrared and Raman Spectra of P9lyatomic Molecules (van Nostrand Reinbold, New York, 1950), p.306.
7.   V.J. Corcoran and R.G. Comeyne, Proc. High Energy/Power Raman Tech. Workshop, R.C. Fukuda and J. Paul, eds., Night Vision and Electro-Optics Laboratory, Ft.Belvoir, Virginia, May 1983.
8.   P. Rabinowitz, A. Stein, R. Brickman and A. Kaldor, Appl. Phys. Letts. 35, 739 (1979).
9.   W.R. Trutna and R.L. Byer, Appl. Opt. 19, 301 (1980).
10.   R. Burnham and N. Djeu, Opt .Letts. 3, 215 (1978).

Efficient Frequency Conversion
of Laser Sources in Nonlinear Crystals

R.L. Byer

Ginzton Laboratory, Stanford University, Stanford, CA 94305, USA

INTRODUCTION

Efficient harmonic conversion of the few well developed solid state laser sources is now an established technology. For example, efficient frequency conversion of Nd:YAG laser to the second, third and fourth harmonic in KD*P crystals is a reliable and commercial technology introduced by Quanta Ray in 1976,

The extension of nonlinear interaction in crystals to parametric tunable oscillators and amplifiers has proceeded more slowly. Although early Nd:YAG pumped LiNbO3 parametric oscillators that were sold as commercial products by Chromatix have continued to operate for more than fifteen years, the apparent difficulty in operating parametric oscillators and the limited quality of available nonlinear crystals has slowed the development of parametric solid state tunable sources.[1]

Recent developments of stable pump laser sources and improved nonlinear crystals may lead to a renewed interest in widely tunable parametric oscillators and in the development of quantum noise limited parametric amplifiers.[2,3]

To date, with few exceptions, the nonlinear media has been oriented and fabricated from bulk grown single crystals. The simultaneous requirements for transparency, excellent optical quality, mechanical strength, nonlinearity and phasematching has limited the number of potentially useful nonlinear materials to a few out of the known class of over 12,000 crystals. The advent of advanced materials synthesis tecyniques such as molecular beam epitaxy now make it feasible to consider designing and constructing synthetic nonlinear materials atom layer by atom layer to fit a particular nonlinear interaction.

HARMONIC GENERATION

The search for and the development of solid state laser sources over the past twenty-five years has resulted in the discovery of many laser systems. However, to date only Ruby, Nd:Glass and Nd:YAG have found widespread use. Based on a quarter century of experience, the prospects for the future development of efficient solid laser sources appears to be limited. Thus efficient frequency extension is an essential aspect of solid state laser engineering if full advantage is to be gained from the few well developed laser sources.

Fortunately, progress has been made in efficient frequency conversion of high peak power solid state laser sources. Energy conversion efficiencies of 50% to the second harmonic and up to 30%

to the third harmonic are now routinely achieved in the conversion of
Q-switched Nd:YAG sources. Even higher conversion efficiencies of up
to 70% are reported for harmonic generation in KD*P of near plane wave
Nd:Glass lasers.[4]

An important step in this direction was the use of Type II second
harmonic generation with two crystals in sequence.[5] Quadrature SHG
allows high harmonic conversion over a wide range of intensities and
thus yields higher overall conversion for non-plane wave intensity and
temporal profile laser beams.

The extension of harmonic generation to high average power laser
sources places other requirements on the nonlinear medium. In addition
to adequate size and nonlinearity, high average power SHG requires
crystals with good thermal properties and a wide temperature phasematching peak. If the laser beam quality is poor, then the nonlinear
crystal must also have a wide acceptance angle. The crystal KTP has
low dispersion which results in a phasematching peak width of 15°C as
compared to 6°C for KD*P. The angular acceptance width is 15 mrad-cm
for KTP and only 5 mrad-cm for KD*P. A further advantage of KTP is a
nonlinear coefficient that is nearly an order of magnitude larger than
that of KD*P. The principal disadvantage of KTP is the small crystal
size and optical quality variations that result from the difficulty
with hydro-thermal growth of KTP. Recent progress in flux growth of
KTP may overcome the crystal size and optical quality limitations.

For harmonic generation of cw laser sources nonlinear crystals
must have a large nonlinear coefficient and be available in adequate
lengths at very low optical loss for harmonic generation internal to
the laser resonator. Soviet scientists have used LiIO3 for efficient
harmonic generation.[6] KTP has been used to efficiently double an
acousto-optic mode locked, cw lamp pumped Nd:YAG source. In China,
Han Kai has used MgO:LiNbO3 for harmonic generation of cw Nd:YAG
lasers.[7] In these experiments, the average laser power levels are on
the order of 1-10 watts of 1.064 µm. At higher average power levels,
care must be taken to engineer for thermal effects such as thermal
lensing in the nonlinear media, thermal offset of the phasematching
peak and thermal induced depolarization of the laser radiation.

The thermal engineering approaches applied successfully to the
slab geometry laser can also be extended to the design of harmonic
generators for high average power doubling. Such techniques are
probably not required up to the 50 W average power level. Using
geometry and thermal averaging approaches such as moving the nonlinear
crystal should allow harmonic conversion of up to 1 kW of laser power.[8]

PARAMETRIC OSCILLATORS AND AMPLIFIERS

Nd:YAG pumped parametric oscillators using LiNbO3 have been well
studied.[1-3] Earlier work showed that LiNbO3 parametric oscillators
operated reliably and provided tunable output over the 1.4-4.0 µm
range. The sources were used successfully for remote sensing
measurements of molecules and of humidity and temperature.[9]

The LiNbO3 parametric oscillator was not developed into a
commercial product because of uncertainty of the crystal supply and
because of the very critical requirements placed on the Nd:YAG pump

laser source. Recent advances in Nd:YAG source engineering, especially the implementation of slab geometry for polarized, low divergence output[10,11] and injection seeding for single axial mode operation[12] have led to renewed interest in the LiNbO$_3$ parametric oscillator as a tunable source.

In addition to LiNbO$_3$, parametric oscillators have recently been demonstrated in AgGaS$_2$ for infrared generation[13] and in Urea for visible and near infrared tuning.[14] AgGaSe$_2$ crystals of the quality required for OPO operation have now been grown and fabricated thus opening the possibility of 3 μm – 15 μm infrared generation. Figures 1,2 & 3 illustrate the tuning ranges of the Urea, AgGaS$_2$ and AgGaSe$_2$ parametric oscillators.

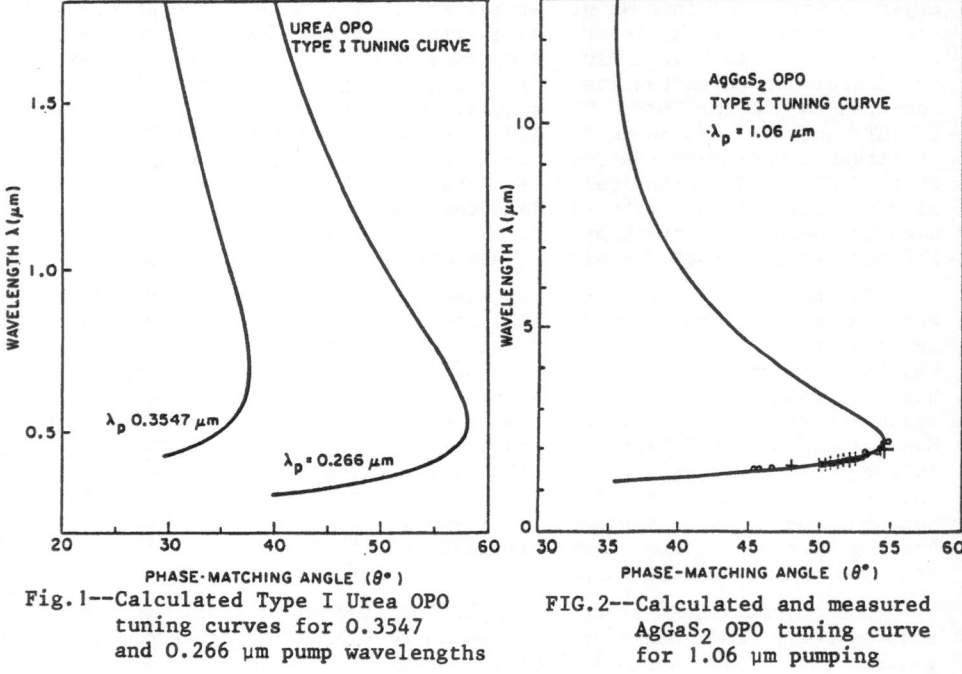

Fig.1--Calculated Type I Urea OPO tuning curves for 0.3547 and 0.266 μm pump wavelengths

FIG.2--Calculated and measured AgGaS$_2$ OPO tuning curve for 1.06 μm pumping

Early work on doubled Nd:YAG pumped LiNbO$_3$ parametric oscillators is also being revisited. The combination of an improved nonlinear material with damage-free MgO:LiNbO$_3$ and narrow linewidth Nd:YAG oscillator and slab amplifiers, open the possibility of cw OPO operation. For example, a 532 nm pumped cw LiNbO$_3$ OPO has a calculated threshold of 60 mW. The OPO is a unique coherent source in that it reproduces the narrow spectral linewidth of the pump laser at the tunable signal and idler output. Experimental work is now in progress to confirm the predicted performance of the cw LiNbO$_3$ OPO. The tuning range of the green pumped OPO is .62 – 3.5 μm.

The parametric amplifier has long been used as a low noise pre-amplifier in the microwave spectral region. Optical parametric amplifiers (OPA) have been studied as power amplifier devices.

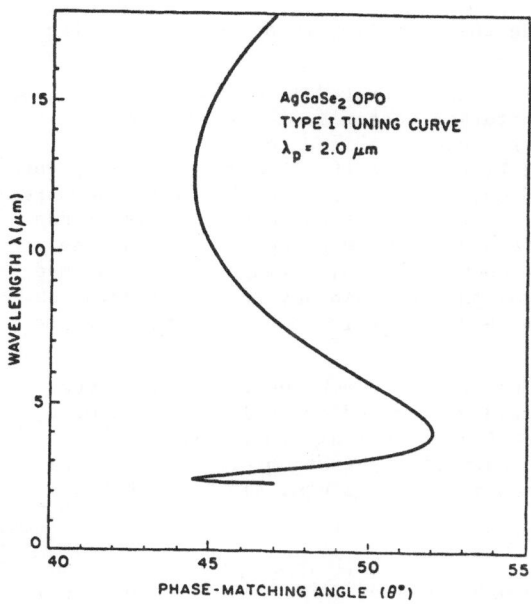

FIG.3—Calculated AgGaSe₂ Type I OPO tuning curve for a 2.0 μm pump wavelength. Operation to 18 μm is projected using a single crystal cut at 48°.

However, the OPA is also a very useful quantum noise-limited optical frequency amplifier. The advantages of the OPA over a dye amplifier are directional, polarized gain over a selectable bandwidth. The OPA thus offers gain without the troublesome effects of parasitic oscill- ations so common in dye amplifiers. Remote sensing applications have not yet required low noise pre-amplifiers as a component of the system. However, future transmitter/receiver systems may benefit from the use of a quantum noise-limited pre-amplifier in the receiver channel.

SYNTHETIC NONLINEAR MEDIA

The design and synthesis of nonlinear media to meet specific interaction requirements is now a possibility due to the recent advances in material synthesis techniques. The ability to construct a nonlinear media with selected transparency, nonlinearity and phase- matching properties is a significant advance over the current practice of optimizing a synthetic or natural bulk crystal by fabrication choices.

Early efforts to synthesize nonlinear media were made in organic nonlinear crystals. The unit cell or acentric organic molecule was designed for optimum nonlinear polarizability by well established organic chemistry synthesis methods. The molecules were then crystallized into a macroscopic array or crystal which in turn was oriented and fabricated for the nonlinear interaction.

Progress in inorganic crystal synthesis has led to the development of integrated optical wave guides in bulk nonlinear media such as LiNbO₃ by diffusion of ions such as protons and titanium. Recent work has centered on the growth of small diameter single crystal fibers to provide the geometry for guiding the radiation to overcome the effects of diffraction.[15]

An alternative approach first demonstrated in China, is to periodically modulate the polarization vector in the ferro-electric crystal LiNbO3 at the spatial frequency required to effect phase-matching.[16] Periodically poled LiNbO3 uses the larger d_{33} nonlinear coefficient and has seventeen times the effective nonlinearity compared to the usual d_{31} interaction. Periodically poled LiNbO3, if it can be successfully grown, also offers a wide acceptance angle and wide phasematching temperature range compared to birefringent phasematched crystals. The ability to control phasematching by spatial modulation of the polarization vector is a very powerful capability which offers wide potential for the material.

Not yet accomplished, but now within reach technically, is the ability to grow periodic modulated or layered crystals. For example, calculations show that a GaAs, GaAℓAs, layered structure is bire-fringent and phasematchable for harmonic generation over the 2 - 8 μm infrared spectral range. The nonlinearity is equivalent to that of GaAs.

CONCLUSION

Nonlinear processes have played an important role in extending the frequency range of available solid state laser sources. With progress in the development of efficient solid state lasers, such as diode array pumped Nd:YAG, nonlinear devices become increasingly the method of choice for frequency extension. The development of synthetic nonlinear media promises to offer an even greater range of devices in the future that are optimized for particular performance goals.

ACKNOWLEDGEMENTS

This work was supported by N.A.S.A. under contract #NAG 1-182

REFERENCES

1. R.L. Byer, "Parametric Oscillators and Nonlinear Materisls" in Nonlinear Optics, eds. P.G. Harper and B.S. Wherrett, Academic Press, p.47 (1977).

2. S. J. Brosnan and R.L. Byer, "Optical Parametric Oscillator Threshold and Linewidth Studies", IEEE Journ. of Quant. Electr. QE-15, p.415 (1979).

3. R.A. Baumgartner and R.L. Byer,"Optical Parametric Amplification", IEEE Journ. of Quant. Electr. QE-15, p.432 (1979).

4. R.S. Craxton "High Efficiency Tripling Schemes for High Power Nd:Glass Lasers", IEEE Journ. of Quant. Electr. QE-17, p.1771 (1981).

5. B.V. Bokut, N.S. Azan, A.T. Malashchenko and Yu.A. Scannikov, "On Specific Features of Second Harmonic Generation in Successively Arranged Nonlinear Crystals", Journ. Prikladnoy Spectroskopii, 37, p.748 (1982).

6. Yu.P. Andreyanov, V.P. Minayev and A.V. Semenov, "An Electro-Optic Shutter for Non-Polarized Radiation in a Laser with Internal Second Harmonic Generation in Crystals Exhibiting an Aperture Effect", Sov. J. Quant. Electron. 10, p.2131 (1983).

7. Han Kai, S.L. Xu, D.C. Pu, L.W. Guo et.al., 14th International Congress on High Speed Photography and Photonics, Moscow, October 1980.

8. David Hon, "High Average Power, Efficient Second Harmonic Generation", in Laser Handbook, ed. by M.L. Stitch, North Holland 1979, p.423.

9. M. Endemann and R.L. Byer, "Simultaneous Remote Measurements of Atmospheric Temperature and Humidity Using a Continuously Tunable IR LIDAR", Applied Optics, 20, p.3211 (1981).

10. W.S. Martin and J.P. Chernoch, "Multiple Internal Reflection Face Pumped Laser", U.S. Patent 3,633 126. 1972.

11. J.M. Eggleston, T.J. Kane, K. Kuhn, J. Unternahrer and R.L. Byer, "The Slab Geometry Laser – Part I : Theory", Journ. Quant. Electr. QE-20, p.289 (1984).

12. Y.K. Park, G. Giuliani and R.L. Byer, "Single Axial Mode Operation of a Q-switched Nd:YAG Oscillator by Injection Seeding", Journ. Quant. Electr. QE-20, p.117 (1984).

13. Y.X. Fan, R.C. Eckardt. R.L. Byer, R.K. Route and R.S. Feigelson, "AgGaS$_2$ Infrared Parametric Oscillator", Appl. Phys. Letts. 45, p.313 (1984).

14. W. Donaldson and C.L. Tang, Applied Physics Letters, 44, p.25 (1984).

15. M. Fejer, G.A. Magel and R.L. Byer, "High Speed,High Resolution Fiber Diameter Variation Measurement System", Appl. Optics, 24, p.2362 (1985).

16. Duan Feng, Nai-Ben Ming, Jing Fen Hong, Yong Shun Yang, Jin Song Zhu, Zhen Yang and Ye Ning Wang, "Enhancement of Second Harmonic Generation in LiNbO$_3$ Crystals with Periodic Laminar Ferro-electric Domains", Appl. Phys. Letts. 37, p.607 (1980).

Infrared Frequency Generation in Chalcopyrite Crystals AgGaS$_2$ and AgGaGe$_2$

R.C. Eckardt

Ginzton Laboratory, Stanford University, Stanford, CA 94305, USA

Both continuously and step tunable nonlinear infrared frequency generation have many potential applications in remote sensing. Nonlinear frequency conversion in the infrared, however, has been limited by the properties of available nonlinear optical materials. The required material characteristics include low absorption, good optical quality, and dispersion and birefingence properties which allow phasematching in addition to nonlinear properties of high resistance to optical damage and a usefull nonlinear optical coefficient. Furthermore, it is essential that the technology exists to reliably grow the material in adequate size and quality.

Several potential materials for nonlinear infrared frequency conversion were identified by the early 1970's. Progress has recently been made in the growth technology of some of these materials. The chalcopyrite materials AgGaS$_2$ and AgGaSe$_2$ have been reliably grown in boule sizes of 2.8-cm diameter by 10-cm length at the Stanford Center for Materials Research. These materials have the potential of nonlinear frequency generation over their combined spectral range of 0.5 μm to 17 μm. Use of AgGaS$_2$ in an optical parametric oscillator tunable from 1.35 to 4 μm, and second harmonic generation of CO$_2$ laser radiation with 14 % energy conversion efficiency in AgGaSe$_2$ are described here.

It was necessary to solve a number of growth related problems to obtain large crystals of good optical quality. Both AgGaS$_2$ and AgGaSe$_2$ are grown in sealed ampoules using the Bridgman-Stockbarger technique. These materials have an anomalous expansion along the crystalline c-axis when cooled from the melting temperature. Therefore it is required to seed boules in the c-direction and to use precision-tapered ampoules to avoid cracking the single crystal when cooling. Large crystals have only been grown with non-stoichiometric composition which results in finely dispersed scattering centers. Post growth heat treatment in the presence of excess Ag$_2$S or Ag$_2$Se is required to remove the scattering centers. Increasingly larger dimensions are being achieved with an ultimate goal of 4-cm-long finished optical crystals. Growth technology is being transferred, and both crystals should soon be available commercially.

Some relevant optical properties of AgGaS$_2$ and AgGaSe$_2$ are listed in Table 1. The spectral ranges over which the two materials are useful for nonlinear frequency conversion are complementary. The sulfide has the advantage of visible transmission for alignment, and it provides phase-

Table I. Properties of AgGaS$_2$ and AgGaSe$_2$

Both are negative uniaxial crystals with $\overline{4}$2m symmetry.

		AgGaS$_2$	AgGaSe
Range of input wavelength over which second harmonic generation can be phase matched.	Type I Type II	1.73 - 11.1 µm 2.5 - 8 µm	3.1 - 13 µm 4.8 - 8.5 µm
Region of transmission		0.5 - 12 µm	0.8 - 17 µm
Range of reported values of nonlinear optical coefficient d$_{36}$ at 10.6 µm in units of 10^{-12} m/V.		12 - 18	33 - 58
Phasematching angle for (Type I) SHG at 10.6 µm.		71.7°	55.0°
Birefringent walkoff angle for SHG pahsematching at 10.6 µm.		0.75°	0.67°
Refractive indices at 10.6 µm	n$_o$ n$_e$	2.347 2.293	2.591 2.558
Refractive indices at 5.3 µm	n$_o$ n$_e$	2.395 2.342	2.614 2.581
Surface damage threshold		13 MW/cm^2 @ 1.06 µm	12 MW/cm^2 @ 10.6 µm

matching and low absorption for tunable frequency generation from the near-infrared to 9 µm. There is a multiphonon absorption in the sulfide that results in a loss of 0.6 per cm at 10 µm. Both the regions of phasematching and of low absorption are shifted to longer wavelengths in the selenide. There is a multiphonon absorption of 0.35 per cm located near 14 µm in AgGaSe$_2$. Absorption at the CO_2 laser wavelength, however, is approximately 0.01 per cm, and this material performs well in second harmonic generation with that source. On the other hand, a AgGaS$_2$ optical parametric oscillator (OPO) will tune to degeneracy with a 1.06 µm pump and provide continuous infrared wavelength tunability, whereas a 1.06-µm pumped AgGaSe$_2$ OPO will not.

We have demonstrated optical parametric oscillation in AgGaS$_2$ pumped with a 1.06-µm, Q-switched Nd:YAG laser. Continuously tunable[2] output between 1.35 and 4 µm was generated by angle tuning of a 2-cm-long crystal. A singly resonant cavity, highly reflecting only at the shorter signal wavelength was used. The tuning range has the potential of extending to 12 µm with the use of longer crystals and optimized resonator cavities. The parametric oscillator, when operated near degeneracy, had a threshold energy of approximately 1 mJ for the 20-ns pump pulse with 0.7-mm beam radius. With increased beam radius, 16 % energy conversion was obtained. Both tuning and energy conversion performance were limited by surface damage at 13 MW/cm^2. With some combination of longer crystals, optimized oscillators, and increased surface damage threshold, it will be possible to achieve the full potential tuning range of this material.

Two-cm-long crystals of AgGaSe$_2$ have been fabricated from recently grown high optical quality material. With such a crystal, 14% internal energy conversion efficiency has been measured in second harmonic generation of a carbon dioxide laser pulse. This measurement was made with a TEA laser pulse consisting of 75-ns gain switched spike that contained 35% of total pulse energy, and nitrogen tail about 650 ns long. Modeling the pulse profile in time and space shows that energy conversion of spike is 26% with 35% peak power conversion and 60% peak intensity conversion. Experimental attempts to convert only the spike were confusing because the pulse clipper that was used shuttered part of the spike and then recovered to transmit a significant portion of the energy in the tail. Conversion efficiency was limited by surface damage which occurred at 12 MW/cm^2 peak intensity or energy fluence of 3 J/cm^2 for the full TEA laser pulse. Still, 14% energy conversion is a useful value, and this level will be increased with longer crystals or increased surface damage threshold.

The growth of AgGaS$_2$ and AgGaSe$_2$ in high quality, large single crystals, and the demonstration of optical parametric oscillation and efficient second harmonic conversion of the carbon-dioxide laser are significant advances in nonlinear infrared frequency conversion. This research is continuing with the construction of a AgGaSe$_2$ OPO, and the growth of these materials in larger sizes. Methods of increasing surface damage threshold are also being investigated. It is likely that these materials will allow the practical, continuously tunable nonlinear frequency conversion from the visible to wavelengths longer than 10 μm in the near furture.

It is a pleasure to acknowledge the individuals who participated in this research. The materials were supplied by Robert Feigelson and Roger Route at the Stanford Center for Materials Reaserch. Optical parametric oscillator research was performed by Fan Yaun Xuan. Jan van der Laan at SRI International collaborated on the second harmonic measurements. Professor Robert Byer guided and coordinated this research.

Nonlinear Optics in Single Crystal Fibers

M. Fejer, J. Nightingale, G. Magel, W. Kozlovsky, T.Y. Fan, and R.L. Byer

Applied Physics Department, Stanford University, Stanford, CA 94305, USA

Since the advent of low loss optical fibers fifteen years ago, considerable research effort has been directed towards the study of nonlinear interactions in fibers. A variety of devices have taken advantage of the combination of transverse confinement and long inter-action lengths available in glass fibers to operate efficiently at relatively low pump powers. Because glasses are inherently centro-symmetric, only third-order nonlinear processes, e.g. Raman[1] and Brillouin[2] scattering, optical Kerr effect,[3] self-phase modulation,[4] or extremely weak quadrupole second order processes[5] are allowed. Thus, the combination of fiber geometry and the second order suscept-ibility of non-centro-symmetric single crystals would open the door to a broad range of nonlinear applications not possible in glass fibers.

The potential of crystal fibers for nonlinear interactions is clear from the theoretical efficiencies of several simple devices.[6] A 25 μm diameter 5 cm long LiNbO3 fiber propagating an HE_{11} mode can double 1.06 μm radiation from a Nd:YAG laser with an efficiency of 0.1% per 1 mW or fifty times the bulk efficiency. Similarly, a parametric oscillator pumped with 532 nm radiation in the same fiber would have a threshold on the order of ten mW. The advantage of the guided wave structure is even more pronounced for interactions involving widely disparate frequencies, e.g. differencing two visible lasers to produce infrared radiation, where the advantage relative to the bulk is several hundred to one.

Fibers useful for device applications must meet fairly stringent quality criteria. In order to maintain phasematching in a parametric process and to minimize scatter losses, the fiber must be a properly oriented single crystal of good optical quality and uniform composition. Ferro-electric fibers must, in addition, be poled, i.e. single domain . Diameter variations can cause phase mis-match, radiation losses and modal coupling . These effects are complicated functions of the core-cladding index difference, the radius of the fiber, and the azimuthal and axial period of the variations. We estimate that diameter variations must be held to less than 0.1 - 1% for typical nonlinear devices.

There are a number of research efforts underway to produce non-linear crystal fibers. Organic crystals grown inside glass capillaries are being investigated by several groups.[7,8] These materials exhibit large nonlinear coefficients (in some cases more than an order of magnitude larger than LiNbO3), and high damage thresholds (comparable to KDP). Crystal properties can be tailored to specific applications by organic synthesis techniques. SHG of a pulsed 1.06 μm laser in a benzil cored fiber was reported several years ago by Nayar.[9] The

interaction was quite inefficient because the second harmonic
radiation was produced in radiation modes of the fiber.

Another approach to nonlinear interactions in fibers is to
embed an unclad glass fiber in a nonlinear crystal. DeShazer has
reported promising results in $LiIO_3$ with this technique.[10] His group
has also grown KDP crystal fibers by an unspecified method.[11]

The growth technique that we have chosen to pursue is miniature
pedestal growth.[12] In this method, the tip of a small rod of the
material to be grown is melted with a CO_2 laser, as shown in Fig. 1.
A seed crystal is dipped into the molten zone, then pulled from the
zone more rapidly than the source rod is fed in. Mass conservation
fixes the diameter reduction as the square root of the velocity ratio.

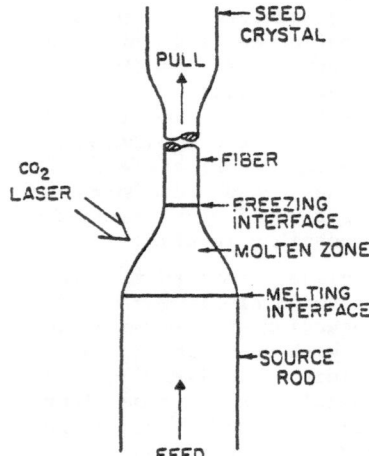

Fig. 1
The laser heated miniature
pedestal growth process.

The abrupt liquid solid transition characteristic of the growth
of crystalline materials is quite different from the viscous draw-
down seen in glass fiber pulling, causing the pedestal growth processes
to be far more sensitive than glass fiber pulling to external perturbations. The growth of fibers suitable for nonlinear applications
therefore, requires a carefully designed apparatus. In particular,
the growth zone must be mechanically and thermally stable and the heat
distribution should be azimuthally symmetrical. The apparatus that we
designed to meet these criteria is shown in Fig. 2.

The novel reflaxicon focussing system simultaneously provides an
azimuthally symmetric heat input and a tight 40 μm focus necessary for
the stable growth of small fibers. A moving belt in the translation
mechanisms slides the fibers through silicon V-groove guides at rates
accurately controlled by regulated d.c. motors. The V-groove guides
prevent motion of the fiber in the plane perpendicular to the growth
axis. A high speed non-contact diameter measurement system described
in detail elsewhere has recently been completed. This device will
allow study of the effect of variations of feed and pull rates and
laser power on the fiber diameter, with the goal of implementing
closed loop control of the fiber cross-section. A block diagram of

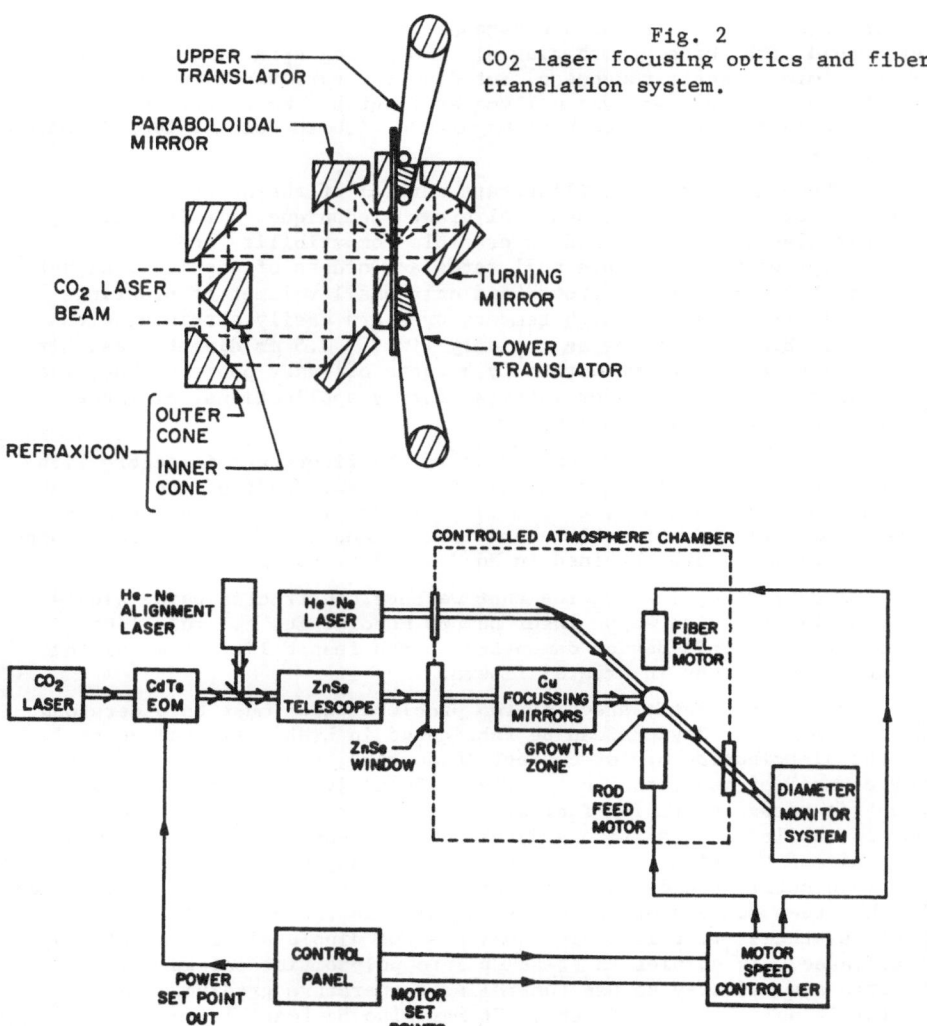

Fig. 2
CO2 laser focusing optics and fiber
translation system.

UPPER TRANSLATOR

PARABOLOIDAL MIRROR

CO2 LASER BEAM

TURNING MIRROR

LOWER TRANSLATOR

REFRAXICON — OUTER CONE / INNER CONE

CONTROLLED ATMOSPHERE CHAMBER

He-Ne ALIGNMENT LASER

He-Ne LASER

FIBER PULL MOTOR

CO2 LASER

CdTe EOM

ZnSe TELESCOPE

Cu FOCUSSING MIRRORS

ZnSe WINDOW

GROWTH ZONE

ROD FEED MOTOR

DIAMETER MONITOR SYSTEM

CONTROL PANEL

POWER SET POINT OUT

MOTOR SET POINTS

MOTOR SPEED CONTROLLER

Fig. 3--Block diagram of the single crystal fiber growth system

the complete system is shown in Fig. 3. Reference 12 gives a more
complete description of the apparatus.

To date we have achieved controlled growth of four materials:
Al_2O_3 , $Cr^{+++}:Al_2O_3$, $Nd^{++}:YAG$ and $LiNbO_3$. Several orientations of
most of these materials have been grown, including both a and c axis
$LiNbO_3$. Fibers with diameters ranging from 20 up to 500 μm have been
grown at rates of 0.5 to 40 mm/min in lengths up to 200 mm. The
necessary CO_2 laser power is typically less than 5 W.

The morphology of the fibers is similar to that of bulk
Czochralski boules of the same orientation. For example <111>

143

Nd:YAG fibers show a rounded hexagonal shape, while <001> LiNbO$_3$ fibers are round with three growth ridges. SEM photographs of the fibers show no micron scale roughness, but diameter variations on the order of 1% rms are observed over millimeter lengths. We expect to reduce these variations by an order of magnitude with the closed loop diameter control system.

These growth results illustrate several of the attractive features of the miniature pedestal growth technique. It is entirely containerless, thereby avoiding crucible compatibility and contamination problems. Feasible pull rates are orders of magnitude higher than in bulk Czochralski growth and only small volumes of starting material are required. High temperatures are easily attained, with the available laser power as the only limit. 0.5 mm diameter sapphire (M.P.2323K) can be grown with only 5 watts of laser power. Thus, the technique is attractive for material survey applications, e.g. new laser host-ion combinations.

The measured propagation losses in the fibers are in accord with theoretical estimates based on the measured amplitude of the diameter variations. For example, a 5 cm long 170 μm diameter ruby fiber had losses of 0.04 dB/cm for 633 nm radiation launched into low order modes. Similar results were obtained in Nd:YAG and LiNbO$_3$ fibers.

The first optical device that we have constructed using single crystal fibers is an argon laser pumped monolithic cw ruby fiber oscillator.[14] This device demonstrates the feasibility of monolithic guided wave devices in crystal fibers.

We are currently studying two problems which must be understood before nonlinear devices can be fabricated in LiNbO$_3$ fibers: control of the distribution of ferro-electric domains, and cladding the fiber for control of modal characteristics. Selective etching and pyro-electric response studies indicate the c-axis fibers grow single domain, while a-axis fibers develop head-to-head domains joined at the axis of the fiber. Both these results can be explained by the thermo-electric fields which are present in the growth zone. Generated by the steep temperature gradients present in the pedestal growth process, these electric fields dominate the dipole-dipole interactions that cause bulk samples to break up into polydomain configurations. Efforts are underway to use controlled temperature gradients to uniformly pole a-axis fibers. It may also be feasible to use periodically varying temperature gradients to form a periodically poled fiber for quasi-phasematching[15] nonlinear interactions.

Techniques for forming low index claddings are also being studied. Such claddings would reduce surface scatter losses and bring the wave-guides closer to single mode operation. Both extruded glass and diffused proton or transition metal claddings have been fabricated and are being tested.

Conclusions

We have designed and built an apparatus to grow single crystal fibers suitable for linear and nonlinear optical applications. Ruby, lithium niobate and Nd:YAG fibers with losses in the 1%/cm range have been grown. An argon pumped monolithic ruby fiber oscillator has been demonstrated.

Future work will proceed in several directions. Studies of cladding and poling a-axis LiNbO3 will continue. The short term goal is demonstration of efficient doubling of 1.06 μm radiation. The first nonlinear device we expect to demonstrate is an electro-optic modulator in c-axis $LiNbO_3$ as the poling problem is already solved for this orientation.

Another thrust of the program will be extending the range of materials grown in fiber form. Two materials to be emphasized are terbium gallium garnet for optical isolators and potassium niobate for doubling gallium arsenide diode lasers.

References

1.   R.H. Stolen, "Fiber Raman Lasers", Fiber and Integrated Optics, 3 (1980).

2.   E. Ippen and R.H. Stolen, "Stimulated Brillouin Scattering in Optical Fibers", Appl. Phys. Letts. 21, 539 (1972).

3.   V. Dziedzic, R.H. Stolen and A. Ashkin, "Optical Kerr Effect in Long Fibers", Appl. Opt. 20, 1403 (1981).

4.   E. Ippen, C. Shank, T. Gustafson, "Self-Phase Modulation of Picosecond Pulses in Optical Fibers", Appl. Phys. Letts. 24, 190 (1974).

5.   Y. Ohmori and Y. Sasaki, "Two Wave Sum Frequency Light Generation in Optical Fibers", IEEE Journ. Quant. Electr. QE-18, 758 (1982).

6.   M. Fejer, R.L. Byer, "Nonlinear Optics in Single Crystal Fibers", to be published.

7.   J. Zyss and J.L. Oudar, "New Organic Molecule Materials for Nonlinear Optics", C.L.E.O., Anaheim, 1984, Session F04.

8.   B. Nayar, D. Smith, C. Yoon and J. Sherwood, "Growth and Assessment of Highly Nonlinear Organic Materials", C.L.E.O., Anaheim, 1984, Session F01.

9.   B. Nayar, in Technical Digest, Topical Meeting on Integrated and Guided Wave Optics, (O.S.A. Washington D.C, 1982, Paper ThA2.

10.   L. deShazer, private communication.

11.   Ibid.

12.   M. Fejer, J. Nightingale, G. Magel and R.L. Byer, "Laser Heated Miniature Pedestal Growth Apparatus for Single Crystal Optical Fibers", Rev. Sci. Instrum. 55, 1791 (1984).

13. M. Fejer, G. Magel and R.L. Byer, "High Speed, High Resolution, · Fiber Diameter Measurement System", Appl.Opt. 24, 2362 (1985).

14. J. Nightingale, R.L. Byer, "A Guided Wave Monolithic Ruby Fiber Laser", to be published.

15. J.D. McMullen, "Optical Parametric Interactions in Isotropic Materials Using a Phase Corrected Stack of Nonlinear Dielectric Plates", J. Appl. Phys. 46, 3076 (1975).

Laser Action of H3 Color Center in Diamond

S.C. Rand and L.G. DeShazer

Hughes Research Laboratories, 3011 Malibu Canyon Road,
Malibu, CA 90265, USA

SUMMARY

We report observation of room temperature laser action of the H3
color center in Type I diamonds. A low-cost green diamond, 2-mm thick,
with polished parallel optical faces was used. The diamond laser was
pumped by a 494-nm laser beam in a longitudinal arrangement and the
laser action occurred from the 18% Fresnel reflections of the diamond
faces. Peak absorption coefficients were near 3 cm^{-1} and gains were
estimated to be 0.09 cm^{-1} for the H3 center. Decay times were sample
dependent and ranged from 15–45 ns. No bleaching of the center was
observed at excitation levels as high as 70 MW/cm^2. This new laser is
expected to have unlimited shelf life, and should be tunable over the
H3 band from 500 to 600 nm.

Intense fluorescence has also been observed in the 400–500 nm
region from N3 centers in other diamond samples. The gain estimated
for N3 centers is 0.009 cm^{-1} but excited state absorption might be a
problem for these centers. Energy transfer between the N3 and H3
centers was observed directly, using a time resolved fluorescence
technique. Single samples containing a mixture of both N3 and H3
centers may provide an unprecedented source of optical radiation
continuously tunable from 400 to 600 nm.

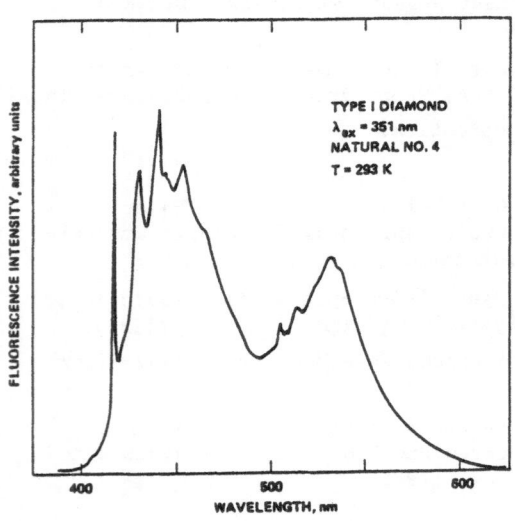

List of Conference Attendees

Frank Allario (804)865-200
M.S. 476
NASA, Langley Research Center
Hampton, VA. 23665.

Peter Banks (415) 497-1830
201 Durand Bldg.
Stanford University
Stanford, CA. 94305

Norman Barnes (505) 667-7732
M.S. J564
Los Alamos National Labs.,
P.O. Box 1663
Los Alamos, N.M. 87545

Leonard Braverman (609) 338-2771
RCA Advanced Technology Labs.,
Front & Cooper Strs. Bldg. 10-4-3
Camden, N.J. 08102

John Brock
T.R.W.
One Space Park
Redondo Beach, CA. 90278

M. Birnbaum
Aerospace Corporation
Los Angeles, CA. 90009

Edward Browell (804) 865-2576
M.S. 401A
NASA Langley Research Center
Hampton, VA. 23665

Stephen Brosnan
Northrop Res. & Tech. Center
1 Research Park
Palos Verdes Peninsula
California 90274

David C. Brown (518) 385-1197
Bldg. 37 ; Rm. 663
General Electric Res. Center
P.O. Box 8
Schenectady, N.Y. 12301

Jack L. Bufton (301) 344-5626
Code 723
NASA Goddard Space Flight Center
Greenbelt, MD. 20711.

Robert L. Byer (415) 497-0226
Ginzton Laboratory
Stanford University
Stanford, CA. 94305

Charles Byvic (804) 865-2000
NASA Langley Research Center
Hampton, VA. 23665

John Caird (415) 422-6159
M.S. 490; P.O. Box 5508; L-490
Lawrence National Laboratories
Livermore, CA. 94550

Robert Curran
NASA Headquarters
Washington D.C. 20546

David G. Dawes (704) 588-2340
Allied Corporation
P.O. Box 31428
Charlotte, N.C. 28231

Larry DeShazer (213) 456-6411
Hughes Research Laboratories
3011 S. Malibu Rd, M.S. RL65
Malibu, CA. 90265

G. Dube
M.S. 2000
General Electric Corporation
P.O. Box 5000
Binghampton, N.Y. 13902

Robert Eckardt
Ginzton Laboratory
Stanford University
Stanford, CA. 94305

Frank Bruni
Materials Progress Corp.
93 Stoney Circle
Santa Rosa, CA. 95401

Arron Budgor (213) 418-4118
Northrop Corporation
230, W. 120 Street
Hawthorne, CA. 90250

John L. Emmett (415) 422-1100
Lawrence National Laboratories
P.O. Box 5508
Livermore, CA. 94550

E.G. ERickson
M.S. 4G07
G.T.E. Sylvania
P.O. Box 7188
Mountain View, CA. 94035

D. Escoe
Space Div. YDMX
Aerospace Corporation
Los Angeles, CA. 90009

Leon Esterowitz (202) 767-3535
Code 6551
Naval Research Laboratory
Washington, D.C. 20375

Michael Ettenberg (609) 734-3149
R.C.A. Laboratories
P.O. Box 432
Princeton, N.J. 08540

J.J. Ewing (206) 827-0460
Spectra Technology
2755 Northup Way
Bellevue, WA. 98004

Martin Fejer (415) 497-1992
Ginzton Laboratory
Stanford University
Stanford, CA. 94305

John Eggleston (206) 827-0460
Spectra Technology
2755 Northup Way
Bellevue, WA. 98004

Richard A. Elliott (503) 645-1121
Oregon Graduate Center
19600 N.W. Walker Road
Beaverton, OR. 97006

Robert Greco
Rm. M-1339
General Electric Company
P.O. Box 8555
Philadelphia. PA. 19101

Steve Guch, Jr. (415) 966-3793
G.T.E. Products
100 Ferguson Drive
Mountain View, CA. 94042

Freeman F. Hall, Jr. (303) 497-6312
NOAA Wave Propagation Laboratory
325 Broadway
Boulder, CO. 80303

Richard L. Herbst (415) 969-3850
Quanta Ray Inc.,
1250 Charleston Road
Mountain View, CA. 94043

Robert Hess (804) 827-2818
M.S. 283
NASA Langley Research Center
Hampton, VA. 23665

Lt. R. Higgins
Space Div. YDMX
Los Angeles Air Force Station
P.O. Box 92960
Los Angeles, CA. 90007

Edward D. Hinkley (213) 354-4321
Jet Propulsion Laboratory - 180/701
California Institute of Technology
4800 Oak Grove Drive
Pasadena, CA. 91109.

Michael Finlan
Rm. M.1339
General Electric Company
P.O. Box 8555
Philadelphia, PA. 19101

William H. Fuller. Jr.
Mail Code 475
NASA Langley Research Center
Hampton, VA. 23665

Gleb Gashurov (201 539-5500
Airtron Div. Litton Industries
200 E. Hanover Avenue
Morris Plains, N.J. 07950

William B. Grant
Jet Propulsion Laboratories
M.S 183-401, Cal. Inst. Techn.
4800 Oak Grove Drive
Pasadena, CA. 91109

Hanna J. Hoffman (415) 493-3311
Lockheed Palo Alto Research Labs.,
Group 95-40, B/201
3251 Hanover Street
Palo Alto, CA. 94304

J. Fred Holmes (503) 645-1121
Dept., of Appl.Phys.& Elect.Engin.
Oregon Graduate Center
19600 N.W. Walker Road
Beaverton, OR. 97005

John F. Holzrichter (415) 422-7454
Lawrence National Laboratories
P.O. Box 5508 ; L-487
Livermore, CA. 94550.

Milton Huffaker (303) 449-8736
Coherent Technologies Inc.,
P.O. Box 7488
Boulder, CO. 80306.

G.T. Hulme
General Electric Corp.
M.S. 2000 ; P.O. Box 5000
Binghampton, N.Y. 13902

Jeff Hutchinson (703) 243-8555
Allied Corporation
1500 Wilson Blvd.,
Arlington, VA. 22209

Ralph Jacobs (408) 946-6080
Spectra Physics
3333 North First Street
San Jose, CA. 95134

Anthony Jalink
M.S. 474
NASA Langley Research Center
Hampton, VA. 23665

Hans Jenssen (617) 253-6878
Room, 13-3154
Massachusetts Institute of Technology
Cambridge, MA. 02139

S.F. Jobes
General Electric Corp
M.S. 2000 ; P.O. Box 5000
Binghampton, N.Y. 13902

Victor Korson (609) 734-2064
R.C.A. Laboratories
P.O. Box 800
Princeton, N.J. 08540

William Krupke (415) 422-5354
L.488, P.O. Box 5508
Lawrence National Laboratories
Livermore, CA. 94550

John Kurmer
M.S. 10-8
R.C.A. Corporation
Front & Cover Streets
Camden, N.J. 08182

T.J. Kane (415) 497-1718
Ginzton Laboratory of Physics
Stanford University
Stanford, CA. 94305.

Dennis Killinger (617) 863-5500
Lincoln Laboratories
M.I.T. 244 Wood Street
Lexington, MASS. 02173

C. Lawrence Korb
GLAS 913
NASA Goddard Space Flight Cntr.
Greenbelt, MD. 20711

Hiroshi Komine   (213) 377-4811
Northrop Res. & Tech. Center
1 Research Park
Palos Verdes,   CA. 90214

M. Kokta
Union Carbide Corp.,
750 S. 32nd Street
Washougal, WA. 98671.

Phil Lacovara   (202) 767-3535
Code 6551
Naval Research Labs.,
Washington D.C. 20375

R.A. Lagno
Airtron Div. Litton Industries
200 East Hanover Avenue
Morris Plains, N.J. 07950

James B. Laudenslager (818) 354-2259
Jet Propulsion Laboratory
Cal.Inst.Tech.   4800 Oak Grove Dr.
Pasadena, CA. 91109.

Kotia Lee   (213) 536-1327
M.S. R1-1169
T.R.W.
One Space Park
Redondo Beach, CA. 90278

Yung Sheng Liu (518) 385-8664
General Electric Research Center
P.O. Box 8, Bldg. KWB-1307
Schenectady, N.Y. 12301

Henry Lum   (415) 965-6544
NASA, Ames Research Center
Moffett Field
Mountain View, CA. 94035

Howard Lowdermilk
Lawrence National Labs.,
P.O. Box 5508 - L-490
Livermore, CA. 94550.

Wayne Lo
General Motors Res.Labs.
G.M. Techn. Center
Warren, MI. 48090

Leonard J. Marabella
T.R.W. M/S R1-1184
One Space Park
Redondo Beach, CA. 90278

Lou Marquet (202)
Defense Advanced Research
        Projects Agency
1400 Wilson Blvd.
Arlington, VA. 22209

Rodney I. McCormick
U.S. Army Research Office
P.O. Box 1221
Durham, N.C. 27701

R. McFarlane   (213) 456-6411
M.S. RL65
Hughes Research Labs.,
3011 Malibu Canyon Road
Malibu, CA. 90265

Robert Menzies (213) 354-3787
Jet Propulsion Labs
Mail Stop 183-401
4800 Oak Grove Drive
Pasadena, CA. 91109

Linn F. Mollenhaur (201)949-5766
Bell Laboratoires
M.S. 4C306
Holmdel, N.J. 07733.

T. Millar
Eastman Kodak Corporation
Rochester, N.Y. 14650

Stephen E. Moody (206) 827-0460
Spectra Technology
2755 Northup Way
Bellevue, WA. 98004

Aram Mooradian
Lincoln Laboratories
M.I.T.   244 Wood Street
Lexington, MA. 02173

Peter Moulton (617) 863-5500
Lincoln Labs., M.I.T.
244 Wood Street
Lexington, MASS. 02173

Richard Nelms
NASA Langley Research Cnere
Hampton, VA. 23665

J. O'Brien
Northrop Res.& Tech.Center
1 Research Park
Palos Verdes Peninsula CA. 90274

J.J. Ozovek
General Electric Corp.
M.S. 2000 ; P.O. Box 5000
Binghampton, N.Y. 13902

Jefferey Paul (703) 644-5310
U.S. Army Night Vision Labs
Fort Belvoir, VA. 22060

Charles R. Philbrick (617) 861-4944
AFGL/LKB
Air Force Geophysics Laboratory
Hanscom Air Force Base, MA.01731

Stephen C. Rand (213) 456-6411
Hughes Research Laboratories
3011 S. Malibu Road
Malibu, CA. 90265

Edward Reed (415) 966-2176
GTE Sylvania
M.S. 4G07 ; P.O. Box 7188
Mountain View, CA. 94039

Robert Rice (314) 232-7567
McDonnell Douglas Astro. Co.
St. Louis, MO, 63166

Joseph F. Rando (408) 946-6080
Spectra Physics Inc.,
3333 North First Street
San Jose, CA. 95134

Mark H. Randles (704) 588-2340
Allied Corporation
P.L. Box 31428
Charlotte, N.C. 28231

Dave Rockwell
Hughes Research Labs, M S. 2264
3011 S. Malibu Road
Malibu, CA. 90265

Arieh Rosenberg (609) 426-2691
RCA Govern. Systems Div.
Astro-Electronics
P.O. Box 800
Princeton, N.J. 08540

Roger Route
Center for Materials Research
Stanford University
Stanford, CA. 94305

C. Roychoudhuri (213) 535-3483
M.S. R1-1078
T.R.W. 1 Space Park
Redondo Beach, CA. 90278

Richard Sam (818) 706-1591
Allied Technologies
31717 La Tienda Drive
Westlake Village, CA. 91362

Kenneth Schepler (513) 255-3804
AFWAL/AADO-1
WPAFB
Dayton, Ohio, 45433

Erhard Schimitschek (619) 225-6224
Naval Ocain Systems Center
Code 8114
271 Catalina Blvd
San Diego , CA. 92152

Richard Schlecht
M.S. 4G07, P.O. Box 7188
G.T.E. Sylvania
Mountain View, CA. 94039

Donald Scifres (408) 946-3483
Spectra Diode Labs. Inc
3333 North First Street
San Jose, CA. 95134

Michael D. Shinn (415) 422-6271
Lawrence National Labs.,
MS. L-490 ; P.L. Box 5508
Livermore, CA. 94035

William Sibley (405) 624-6501
101 Whitehurst Hall
Oklahoma State University
Stillwater, OK. 74078

Martin Sokoloski (202) 453-2864
NASA Headquarters; Code RC
Washington, D.C. 20546

Walter Sooy
Lawrence National Labs
P.O. Box 808
Livermore, CA. 94035

William Streiffer
Physics & Astronomy Dept
Rm. 115, Univ. of N.M.
lbuquerque, N.M. 87106

W. Tiffaney
Coherent General
Palo Alto, CA.

Rohert Uhrin (201) 539-5500
Airtron Div. Litton Industries
200 East Hanover Avenue
Morris Plains, N.J. 07950

Edward Uthe (415) 859-4667
S.R.I. International
333 Ravenswood Ave
Menlo Park, CA. 94025

John Walling (201) 560-1750
Allied Chemical
P.O. Box 4901
Mt. Bethel , N.J. 07060

Rohert Warren (704) 588-2340
Allied Corporation
Synthetic Crystal Products Div
P.O. Box 31428
Charlotte, N.C. 28231

F. Way (505) 242-0393
Specron Div. Labs. Inc.,
2650 Yale Blvd. S.E.
Albuquerque, N.M. 87106

Thomas D. Wilkerson (301) 454-5401
Inst. Phys. Sci. & Tech.
University of Maryland
College Park, MD. 20742

Y.W. Yin
General Electric

Index of Contributors

Springer Series in Optical Sciences

Editorial Board: J. M. Enoch, D. L. MacAdam, A. L. Schawlow, K. Shimoda, T. Tamir

Springer-Verlag
Berlin Heidelberg
New York Tokyo